Public Policy Issues in Resource Management

Electric POWER

and the Future of the Pacific Northwest

Kai N. Lee and Donna Lee Klemka,
with Marion E. Marts

FOREWORD BY SENATOR HENRY M. JACKSON

UNIVERSITY OF WASHINGTON PRESS
Seattle and London

Printed in the United States of America

Library of Congress Cataloging in Publication Data

Lee, Kai N
Electric power and the future of the Pacific Northwest.

(Public policy issues in resource management; 6)
Bibliography: p.
Includes index.
1. Electric utilities--Northwest, Pacific.
I. Klemka, Donna Lee, joint author. II. Marts, Marion Ernest, 1915- joint author. III. Title. IV. Series.
HD9685.U6A1944 333.79'32'09795 79-6762
ISBN 0-295-95720-4

Contents

Figures

Tables

Foreword

We are at a major turning point in the history of electric power development in the Pacific Northwest. In the 1930s great decisions were made to develop the power potential of the Columbia River. At the time the Columbia seemed a virtually limitless source of reliable and inexpensive energy. When the Grand Coulee and Bonneville Dams were built, it was anticipated that those dams alone would supply the then projected power needs of the Northwest for the subsequent twenty years.

J.D. Ross, the first Administrator of the Bonneville Power Administration and a former Superintendent of Seattle City Light, described the Columbia as "a coal mine that never thins out. It is an oil well that never runs dry. The Columbia will flow through the Bonneville and Grand Coulee Dams...as long as rain falls and water flows downhill to the sea."

When those words were spoken, the biggest problem facing the Northwest was finding a market for the vast surplus supplies of power which those dams made available. What a different world we live in today!

Now the Columbia River power system is largely developed. The river is vast, but it is finite; and though it continues to flow through the great dams of the Columbia, it is no longer sufficient to fully meet the needs of the region.

Today the traditional assumptions of power planning are all reversed. Present supply is limited. New sources of supply are becoming harder and harder to plan, finance, and construct; and the only thing that is limitless in the power development field any more seems to be the price at which new generating resources come on line.

The challenges of regional power have mounted steadily over the past seven years. In 1973, the Bonneville Power Administration ceased selling power to private utilities. Subsequently, major industrial customers of BPA were advised that their power contracts would not be renewed as they began expiring beginning in 1981 and "preference customers" of BPA--publicly owned utility systems--were given "notices of insufficiency" which meant that after 1983 Bonneville would not guarantee to supply the full loads of those agencies.

These developments made it clear several years ago that major legislative action would be required to meet the emerging problems of regional power supply. To initiate regional discussion, in the autumn of 1977 I introduced a regional power bill drafted by the Pacific Northwest Utilities Conference Committee. It represented the best efforts of the utilities and industrial users of electric power in the Northwest to come to grips with the problems of the region. Extensive regional discussion and debate followed, and the Senate Energy Committee held four days of hearings in the region during April and May of 1978.

I introduced legislation in August 1978, on behalf of myself and the five other Northwest senators, which reflected many of the suggestions received in the hearings. Although time did not permit full consideration of this bill in the 95th Congress, the Senate passed an amended version of this legislation on August 3, 1979, and I hope that the House of Representatives will enact companion legislation before the end of the year.

These Congressional actions will preserve and enhance the best and most reliable utility system in the nation. The challenge and opportunity will shortly return to the logical center of the action: the Pacific Northwest. The Northwest has long benefited from a diverse and pluralistic utility structure. Public power developed to maturity in our region, and its traditions of public service and economic development run deep in the national grain. We enjoy--and will continue to enjoy through cooperative regional planning and financing programs--the lowest power rates in the country.

As we move toward the future, conservation and regional planning will assume a new importance. Both require an expanded citizen awareness and a broader public vision. I am confident that the people of the Northwest will lead the nation, as they have during the long, successful development of hydroelectric power.

Yet the sheer complexity of the issues at stake in regional power development poses a considerable barrier to public understanding and confidence. The drawn-out debate over the regional legislation has touched upon many matters of vital economic, social, and environmental concern to the people of the Northwest.

Thus I welcome this analysis of the vital role of electric power in our region, and the changes which have reshaped that role during the last decade. Professor Lee, Ms. Klemka, and Dean Marts have made an important contribution, not only to scholarly knowledge, but to the citizen concerned to understand the indispensable complications of energy in our public life. While I may differ with some of the judgments of these authors, their portrait of electric power in the Pacific Northwest captures the rich texture of the opportunities and challenges that we face.

Henry M. Jackson

Washington, D.C.
August 1979

Preface

Since the early 1960s, the Graduate School of Public Affairs has, with the assistance of faculty from a number of schools and colleges of the University, sponsored a natural resources public policy seminar. These seminars have provided opportunities for interdisciplinary examination of critical natural resources and environmental policy issues by faculty members and graduate students. This collaboration has resulted in research projects, new courses and programs, advice to governmental and industry leaders, and the publication of the series of books of which the present volume is the latest. In all of these activities, whether formally associated with the seminar or deriving from the associations and interests the seminar stimulated, the central theme has been to mobilize the collective intellectual attention of both academic disciplines and professional programs on resource and environmental public policy issues of current concern. In this volume, the sixth in the series, the authors return to a theme always central to natural resources and environmental policy issues in the Pacific Northwest.

For most of the twentieth century electric power has been a vital and strategic resource in the Pacific Northwest. The abundance of electric energy for industrial, agricultural, and urban development has shaped this region and its relations with the national and global economy in important ways. So rapidly has economic growth proceeded that the hydroelectric power latent in the Northwest's rivers has reached nearly full development in less than half a century.

Thus, the region--like the industrial world of which it is part--stands at a crossroads: we must plot both wisely and swiftly a new course for energy use and development, one which can conserve the wealth and progress won thus far, while sensibly engaging with the limits of nature and the quality of life, democratically defined. The energy crisis, then, has come to the Pacific Northwest, but with a distinctive regional agenda: it is electric power which leads the list, rather than petroleum; cost, rather than immediate supply, is the dominant consideration; and the choices available to the people of the Northwest are still promising, rather than painfully stark.

This study maps the history of electric power in the Pacific Northwest, and discusses the complex intertwining of technology, economics, institutional arrangements, and political choice now ahead. Decisions made and implemented in the next four years--no later than mid-1983--will shape the future of the region for a generation or more, just as the decisions leading to the Bonneville Project Act of 1937 and subsequent federal power programs have shaped regional electric power development in the Pacific Northwest for the past forty years.

Work on this book began in 1974 with a faculty and graduate seminar on electric power in the Northwest, organized by Professor Lee. This seminar continued in somewhat different form as Natural Resources Public Policy Seminar initiated in 1963 by the Graduate School of Public Affairs. The seminar led to a study, undertaken by Professors Lee and Marts, sponsored by the Washington Water Research Center at Washington State University, and the Office of Water Research and Technology of the U.S. Department of the Interior. Additional support came from the Northwest Energy Policy Project in 1976, the U.S. Office of Environmental Education, Department of Health, Education and Welfare in 1977-78, and the Institute for Environmental Studies. The authors and the Graduate School of Public Affairs gratefully acknowledge these sponsors' support; it has been instrumental in the successful completion of this volume.

The study has also produced direct educational benefits. Six graduate students--in political science, civil engineering, urban planning, geography, and communications--have participated in work reported here. Donna Klemka's masters degree project in geography was the most comprehensive contribution, and she earned her title as co-author early in her work on this project. Preliminary versions of this work were presented in reports written for funding agencies, and discussed in seminars at the Graduate School of Public Affairs, the Institute for Environmental Studies, the Energy and Resources Program of the University of California, and before committees of state legislators and the United States House of Representatives.

The authors wish to acknowledge the advice and criticism of a number of academic colleagues and persons working on regional power issues: Professors William A. Brewer, David J. Olson, Margaret A. Levi, and Stuart A. Scheingold of the University of Washington; Professor J. Serge Taylor of the University of Michigan and Professor Dorothy Nelkin of Cornell University; Mr. Edward W. Sheets of the U.S. Senate staff; the staff of the Bonneville Power Administration, especially former administrator Donald Paul Hodel and Mr. George Reich; Mr. Ken Billington, Dr. Terry R. Lash, Mr. Robert O. Marritz, Mr. Eric Redman and Mr. Dan Seligman. Professor Gordon H. Orians and Mrs. V. Crane Wright of the Institute for Environmental Studies have provided essential support and encouragement; Professor Lee wishes to acknowledge the support and advice of Dana Jeffrey Lee and Professor Donald R. Matthews. The authors' statements and judgments are, of course, their own.

Mrs. Barbara C. Peterson supervised production of the manuscript, and Ms. Lane Morgan edited the final version. Permission to reproduce illustrations from the sources noted in the text is gratefully acknowledged.

Brewster C. Denny, Dean
Graduate School of Public Affairs

Robert H. Pealy, Director
Institute of Governmental Research

Seattle, August 1979

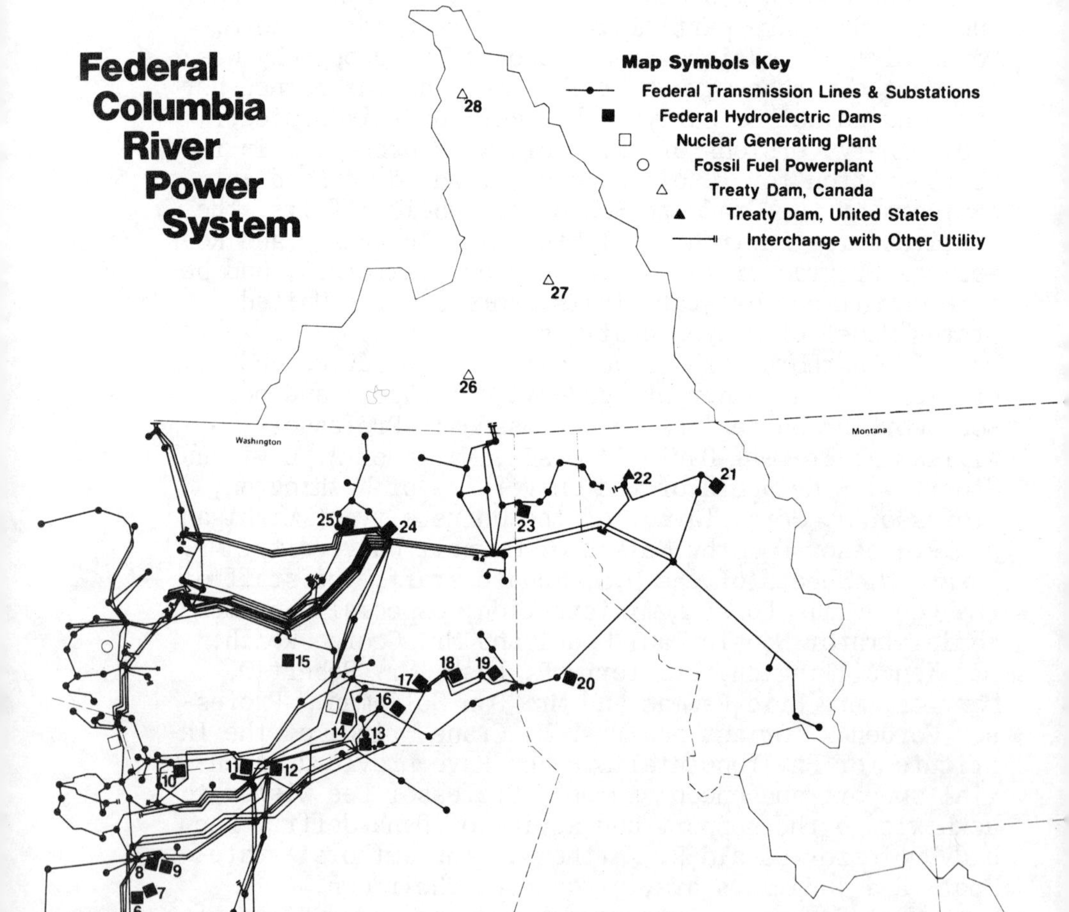
Federal Columbia River Power System
Map Symbols Key
Federal Transmission Lines & Substations
Federal Hydroelectric Dams
Nuclear Generating Plant
Fossil Fuel Powerplant
Treaty Dam, Canada
Treaty Dam, United States
Interchange with Other Utility
Washington
Montana
Oregon
Idaho

ELECTRIC POWER and the Future of the Pacific Northwest

CHAPTER I
Generation at the Turning Point

> Electricity in Napoleon's day had about the same importance as, in the days of Tiberius, could have been ascribed to Christianity. It is gradually becoming apparent that the general innervation of the world by electricity is more fraught with consequences, more capable of modifying life in the near future, than all of the so-called "political" events which have happened from Ampere's day to the present time.[1]
>
> --Paul Valery

From the dark green of its western coasts to the majestic crumpled canvas of its arid interior, the Pacific Northwest is a land of grandeur and opportunity. Its coastal tribes lived in a fertile environment, amid abundance virtually unknown among prehistoric peoples. Today, the region flourishes, discovered by the national media as a place of exceptional "livability." Construction demand is strong and so is the need for lumber from the evergreen Northwest. The dramatic rise in world food prices during the 1970s has brought prosperity to agriculture in the Columbia River Basin. Seattle's Boeing Company is handily maintaining its leading position in commercial aircraft as orders for airplanes pour in from around the world. And the trans-Alaska pipeline, together with blossoming trade around the Pacific Rim, brings a torrent of commerce to the region's ports. In lifestyle, resources, and economic vitality the Pacific

Northwest rivals the fabled Sunbelt as an apotheosis of the American Dream.

That dream, as the troubled 1970s make manifest, is riven by contradictory claims: growth versus quality of life; prosperity versus inflation; energy versus environment; "looking out for Number One" versus "common cause." These claims compete to put their stamp on the future of the Pacific Northwest. A key and controlling element of that future is electric energy.

THE SETTING

Since Edison's carbon-filament bulbs lit up lower Manhattan in 1882,[2] electricity has been a hallmark of modern civilization. Electric energy has been fundamental to the development of the Northwest: the taming of the Columbia River provided many benefits, but the most dramatic has been the hydropower produced at the river's mighty dams to power industry and agriculture, to light, heat and make convenient rural, urban, and (inevitably) suburban life.

"Roll on, Columbia, roll on," Woody Guthrie wrote, "Your power is turning the darkness to dawn,"

> And far up the river is Grand Coulee Dam,
> The mightiest thing ever built by a man,
> To run the great factories and water the land,
> It's roll on, Columbia, roll on.[3]

Grand Coulee is no longer so far from the civilization it helped to build, and the roll of the Columbia has become the stately pulsing of reservoirs impounded and controlled by dams. The awesome energy of the river has been harnessed. What does still roll on is the river of economic growth, bringing with it questions that no longer have clear answers: How much additional energy is needed to sustain the economic prosperity of the Northwest? How shall the regional economy be coordinated with that of the nation? Who shall decide, and how? As the conditions of its flourishing suggest, the Northwest is no longer isolated, and the questions that cloud its future are ones that loom over national policy, world trade, even industrial civilization itself, as we wrestle with the challenges of energy.

In 1975 the region's leading energy official,

Bonneville Power Administrator Don Hodel, pointed out that within 20 years the Northwest would need "jobs for more than 800,000...young people--the children who are already here."[4] What would it take? The conventional wisdom of the business community is that employment grows with energy use: electric "loads will increase nearly three times"[5] by 1995; if not, those jobs would not be there when needed. Less than two years later, a government study forced by an environmental lawsuit declared, "one cannot conclude...that growth in GNP 'requires' energy consumption."[6] How much electricity does the Northwest require to maintain economic prosperity? Can the region continue to nurture its quality of life? The answers are in doubt, and the time to decide approaches relentlessly.

A third of the aluminum produced in the United States comes from the Pacific Northwest, aluminum for airplanes, soda cans, and the lighter auto bodies that will help save petroleum in the 1980s. But aluminum is made with electricity, lots of it. In fiscal 1978, the six aluminum manufacturers of the Northwest purchased nearly 24 billion kilowatt-hours of electric power, more than a third of the output of the Federal Columbia River Power System.[7] Is this a sound use of the region's increasingly scarce and valuable hydroelectric power? The industry employs relatively few people. But the importance of aluminum to the U.S. economy, and even to national defense, is undeniable. How much power for aluminum is enough--and who should decide?

"Who should decide?" has become an international question, as the industrialized world struggles with the ever thornier problems of energy. The National Energy Act of 1978 was mired for nearly two years in Congress, and the development of a coherent Western European energy policy stumbled at the outset and fumbles still.[8]

The emergence of a comprehensive institutional structure to manage the regional electric power system in the Northwest could thus be an important model for this nation and others. David Freeman, now chairman of the Tennessee Valley Authority, wrote not long ago:

> The present industry structure is clearly inadequate. Region-wide generation and transmission companies...would be a tremendous im-

> provement. These companies would be a natural outgrowth of the present power pool arrangements where coordination takes place through coordinating committees. The difference is that a full-time corporate management...would have the responsibility and the authority to plan, build and operate a region-wide grid. Sites would be selected on the basis of the best land-use planning for the region as a whole. Duplication would be eliminated and reliability of service improved.[9]

The Pacific Northwest and the Tennessee Valley have long led the nation in the sort of regional planning Freeman advocates. Among the opportunities to be grasped in the Northwest's electric power debate, none is more significant for the region and the nation than the reconstruction of an institutional structure able to run a multistate electric energy development and conservation program. But to accomplish this, a generation of suspicion, competition, and uneasy truce must be transcended --a task of high political art.

Above all, there is cost. Today, the average monthly electric bill in Seattle is far below the national average--at $7 for 500 kilowatt-hours, one-sixth the cost of service in New York City.[10] Correspondingly, residential electricity use is higher than in other parts of the country--four times as high in Seattle as in New York.[11]

The regional electric power supply system, shaped by the geography of its hydropower, is integrated on a multi-state scale rare in the United States. About three-fourths of the region's electricity is supplied by hydropower--a resource of just over twelve million kilowatts, enough energy to serve a dozen cities the size of Seattle. Most of the remaining power is generated at three coal-fired projects and two nuclear plants.[12] Half the electric power in the Pacific Northwest is produced at federal dams and marketed by the Bonneville Power Administration, an agency of the U.S. Department of Energy headquartered in Portland, Oregon. In fiscal year 1978, 76.5 million megawatt-hours of power were delivered by BPA to 160 customers over a 12,500 mile regional transmission network; the agency received revenues of $268 million in return.[13]

The rough average price implicit in these numbers is three-tenths of a cent per kilowatt-hour--approximately one-seventh the typical cost of bulk power elsewhere in the nation.

CHALLENGES

But the abundant energy of the Columbia and other Northwest rivers is now largely developed. The future holds many problems and no panaceas. There are three intertwined dimensions: technology, economics, and institutional arrangements.

The technical requisites of expanding power supply are complex for two general reasons. First, the technologies now to be harnessed--coal, nuclear, and renewables like windpower and eventually solar--are themselves intricate, systemic undertakings unfamiliar in the Northwest. Second, these new supplies must be smoothly merged into the established pattern of regionwide interdependence that is the legacy of federally sponsored hydropower. The complex fabric of constraints and opportunities affects utility operations, choices of new power supply, and environmental quality. Consider the implications for hydro.

The damming of the Columbia River dramatically changed its natural cycle of flow, controlling floods and revising the ecological ground rules by which its wildlife live. Chinook salmon once ranged 1,200 miles, to the headwaters of the Columbia; today, they are stopped by the walls of Chief Joseph Dam on the Columbia and Hell's Canyon on the Snake, each 600 miles from the Pacific.[14] Furthermore, the enormous generating capacity of the dams on the Columbia and its tributaries, capable of handling peak loads of over 24,000 megawatts, has the potential for even larger, more uncertain environmental implications in the future.[15] Yet the extraordinarily low cost of generating Columbia River hydro promises an unusually economical solution to one of the traditional problems of electrical technology.

Electricity cannot be stored. It must be generated as it is needed, and the need fluctuates each day. Using nuclear or fossil-fueled plants, expensive generating equipment must lie idle much of the time. Where it is available, hydropower is a highly desirable source of electric power for meeting the varying portion of demand, because the output of dams is easily adjusted. In the Northwest, moreover, the existing network of hydro projects has been built at costs far below those that prevail today. The economic utilization of hydro means, however, that variations in the flow through the dams will grow increasingly large, as the overall demand mounts in the future, and the size of daily fluctuations grows with it. The fluctuations will lead to surges in the water level at the foot of Grand Coulee Dam at least fifteen feet high in 1985, bringing environmental changes that remain poorly understood.[16]

The hydro resource is largely harnessed. Further developments will raise the peak generating capability of the hydro system, but existing dams already capture most of the available energy. The alternatives fall into two broad classes: thermal power, mainly fueled by coal or nuclear energy; and various forms of unconventional generation.

The unconventional sources make use of renewable resources, such as wind, or seek to improve the efficiency of fuel use through co-generation, making electric power as a by-product of industrial process heat. Unconventional sources tend to be less constant in their output than technologies specifically developed to provide electric power. Therefore, merging these sources with the existing system of supply poses problems of technical coordination. In this, however, the Pacific Northwest enjoys two advantages. The first is the flexibility of the hydro, which can be used to enhance the reliability of unconventional supplies. Second, BPA's region-wide transmission grid provides economical dispatch of power generated anywhere within the region to points of consumption.

Conventional thermal generation today relies principally on coal-fired plants, mostly located on the eastern fringe of the region, in Montana and Wyoming. Like much of the western coal on which the national energy program has pinned its hopes, these resources are far costlier than existing hydropower, in both economic and environmental terms. Air pollution and the scarring of massive tracts of land, together with the boom-town disruptions of rapid industrialization, are some of the harsh costs of additional electricity that distant consumers do not often see.

Nuclear energy is still surrounded by controversy. But at the confluence of the Columbia and Snake Rivers, in the Tri-Cities area of south-central Washington State, it has found a secure niche. There, in communities nourished by the atomic energy establishment since the Second World War, support for nuclear energy is strong, even feisty. Elsewhere, argument swirls, and deadlock is the order of the day.

Whether fueled by coal or nuclear power, thermal projects are large--commensurate in some cases with the largest Columbia River dams. Like the dams, thermal facilities are of regional scale: each plant serves customers in several states. Thermal power thus deepens the interdependence inherited from the hydroelectric period. The principal difference is that large power projects are no longer built by the federal government, which constructed most of the dams. As we shall see, a central concern of the region's utilities has been to find ways to cushion the large risks implicit in interdependence, using other federal means where possible.

In the short run the main hope for regional power management is conservation. Wiser, more parsimonious use of electric energy holds the only prospect of maintaining low-cost power, since this would stretch further the benefits of Columbia River hydro. The technical means for conserving energy are relatively straightforward, although the implementation of conservation measures on the scale envisaged requires a good deal of social coordination and change in public attitudes. Conservation thus seems likely to require regulatory measures enacted by state, local, and federal governments. Like other decentralized technological changes, such as the installation of catalytic converters on automobiles, regulation can create institutional problems in its own right.

More seriously still, conservation highlights the problems of allocating the existing low-cost federal hydro resource. Although costs of generation are rising throughout the Pacific Northwest, they are doing so unevenly, and the customers denied access to federal hydro are seeing their rates mount the most swiftly. In the best-known case, residents in Portland, Oregon, pay nearly three times as much as their neighbors in Clark County, Washington, who have legally protected access to federal hydropower.[17] Allocation has become an economic sore point, and the pain has reached the body politic. Conservation, once a solution advocated by a minority of environmentalists, has become universally touted as a necessity; it is a necessity supplied only in the breach thus far.

The complexity of the regional power supply system is compounded by the large economic implications of technological choice. There is substantial risk of high costs from all forms of action--and inaction. Conservation, the least uncertain from a technical standpoint, requires discipline and coordination that is novel and instinctively unwelcome in a society accustomed to equating growth with progress. Utilities must make costly decisions based on estimates of future power demand, and it is hard to predict the success of conservation efforts. If not enough is conserved, and demand grows, curtailments could cause economic hardship to industries with large appetites for power. Building more power supply than is needed, on the other hand, is expensive as well. If surplus power cannot be exported, its capital costs may have to be borne by consumers throughout the region in the form of higher prices. Most of all, the benefits of slowing demand must be clearly divided beforehand, in order to provide individual utilities with the motivation to undertake meaningful conservation measures.

The economic challenge of financing thermal plants poses, in some ways, the reverse problem. There are large costs at the outset of thermal plant development, costs which must be divided up before the project is launched. Many of the benefits of region-scale thermal development are shared broadly, while costs must be assigned to individual utilities in a legally binding, precise fashion. Accordingly, a mechanism for pooling the utilities' ability to generate capital is needed, as well as a means for sharing the risks of uncertain costs and technologies untried at this scale.

That the costs of conservation and the benefits of thermal plant development are not yet precisely known is in part a reflection of the long-term nature of these undertakings. In the short run, controversies over individual plants and the allocation of hydropower dominate the disputes between utilities, government, and citizen activists.

Reasonable people disagree on how much additional power is needed and how quickly. There are now thirteen thermal power projects in operation or under construction in the Pacific Northwest; nine are nuclear, four coal-fired. The utilities' West Group Forecast, the official industry projection, calls for a fourteenth thermal project--called simply "Plant A"--to go into service in July 1989.[18] The West Group projection of generating resources shows twelve further thermal power stations--Plants B through M--added to the regional supply by mid-1998. This building program would more than double the Northwest's power production capacity by the end of the century. In 1977 the Natural Resources Defense Council, an environmental law firm with offices in northern California, estimated that with a strong conservation program, only eight thermal plants would be needed by 1995-96--all of them currently under construction or in operation. NRDC suggested that six projects could be cancelled or deferred, and that no new starts were necessary.[19] The West Group Forecast and the NRDC Alternative Scenario vary by 70 percent in their projected needs, the primary difference being the assumed intensity and success of conservation.

Meanwhile, the cost of thermal generation continues to grow. The rapidly rising construction costs of the 1970s have cast a dark shadow over thermal plant construction. Paradoxically, skyrocketing inflation has meant that there is nothing to be gained from waiting to learn by others' experiences. Instead, economic pressure favors rapid construction, lest power plant costs mount even further. The five nuclear projects of the Washington Public Power Supply System have jumped in cost, from $4 billion at the time of their original announcement, to over $11 billion today; none of the plants has yet been completed. This escalation has been accompanied by delays in the projects averaging more than two years per power plant--a total of more than fourteen years' slippage overall.[20] The cost increase in the WPPSS plants exceeds in dollar value the capital invested in the Federal Columbia River Power System, comprising the costs allocated to electric power production in the federal dams and Bonneville's transmission grid.[21]

Sizable as these stakes are, the value of the hydropower to the region is substantially larger. For the true worth of the hydropower is gauged, not by the number of federal dollars spent on it in the past, but by the costs necessary to duplicate that system in the present, were it not already there. The 76.5 billion kilowatt hours of electric power generated by the federal system in fiscal 1978 would cost well over a billion dollars were they generated in new thermal plants. This should be compared with the $268 million of BPA revenues earned through power sales that year.[22] Differences of this magnitude are politically important.

The technical and economic texture of regional electric power in the Pacific Northwest would make the task of managing such a system intricate at best. But the greatest challenge by far is the winning of regional and national assent to a new institutional structure for management. The complications of the institutional structure reflect not only the challenges of the present and future, but conflicts and divisions inherited from the past. The key question is how the Bonneville Project Act, signed into law by President Franklin D. Roosevelt in 1937, can be rewritten to provide a structure for regional power planning and operations. This structure should be appropriate to a developed hydro system and a prosperous economy, though one subject increasingly to cost and environmental constraints in the generation of electricity.

From its origins, Northwest electrical energy development has reflected a division between privately owned utilities and publicly owned systems run by governmental authority at the municipal or county level or by rural cooperatives. Federal power development accorded preference to publicly owned utilities in the allocation of hydropower. The reservation of low-cost federal energy for such entities was a powerful spur to the formation of public power agencies in the 1930s and '40s. Yet until the federal hydro resource was fully developed, there was usually sufficient power generated in the federal system to meet substantial demands from privately owned utilities as well. This is no longer true; as a result customers served by investor-owned utilities have seen their power rates mount rapidly in the 1970s as the private companies have had to build their own facilities.

The public power movement led to a highly fragmented organizational framework, as numerous small public agencies were formed during the 1930s and '40s, drawing power from the federal system and distributing it to ultimate consumers. Today, the Bonneville Power Administration serves 116 public agency customers in the Pacific Northwest, many wholly dependent on federal supplies. Moreover, the multi-state span of the regional electric power system means that state jurisdictional boundaries do not encourage regional cooperation.

Most important, revision of the institutional structure must now reflect a shift in the political composition of the Pacific Northwest. Throughout the hydroelectric era, electric power was so low in cost that policy in this area was viewed as the exclusive preserve of the utilities. The ideological thrust of the public power movement relied implicitly on the indefinite availability of ever-larger quantities of cheap power to advance its ends. But electric power prices are likely to rise more rapidly than inflation for the foreseeable future. Moreover, the high costs of further power development are not only economic but environmental and social as well. This means that concern over rising prices will be sustained and elaborated by controversy over the indirect non-economic costs of electric power generation. Regional power planning is no longer only a utility problem.

Institutional revision is thus politically charged. Yet the complexities of the issues inhibit mass participation and broad public understanding. The political pressures that have been mobilized so far emphasize short-run economic concerns, especially conservation and the allocation of federal hydropower. Despite the intensity of the political cross-currents, attempts to reach compromise solutions have been restricted to a small elite group of knowledgeable government officials, utility executives and representatives of industry and interest groups. Negotiation has been slow, difficult, and often subject to the misunderstandings and frictions of personality conflicts.

The issues divide the interests in many ways. There

are, as a consequence, some unexpected alliances. In 1978, for example, the aluminum industry agreed to subsidize electric rates for residential customers of private utilities, an arrangement ratified by the U.S. Senate a year later. City officials in Portland, to take another example, have twice within the decade led unsuccessful attempts to disenfranchise the city's two privately owned utilities. But in 1978 the utilities briefly joined Portland as plaintiffs in a suit against BPA, claiming a share of low-cost federal hydro. Power politics makes for strange bedfellows, and these coalitions often inhibit the design of administrative structures capable of managing regional electric power problems. The leading voices for regional conservation policies--Oregon's Department of Energy and the City of Seattle--are separated by their contrary positions on federal hydropower allocation. Moreover, there has been a fear, shared by many for different reasons, of dominance by the Bonneville Power Administration. Yet Bonneville remains the leading regional actor, as it has been for a generation.

RECOMMENDATIONS

The sense of urgency in the regional power debate reflects in part the national focus on energy policy as the central symbol of the deadlocked democracy of the 1970s. Indeed, the political problems besetting electric power planning in the Northwest are often discussed as if they were coterminous with the energy problem generally. Yet national concern is centered on petroleum and, since Three Mile Island, on nuclear safety. Energy is not a single problem, either at national or regional scale. Moreover, energy is not a temporary problem.[23] The scale and tempo of change in the Pacific Northwest electric energy situation is one that will evolve under conditions of high uncertainty for the rest of this century, at least.

Like many other issues in American politics, then, energy must compete for the attention of a divided public. The neo-conservative suspicion of government meddling, increasingly popular in the tax revolts of the late '70s, means that options for working on any social problem in the public sphere may be sharply limited.

The practical necessity, therefore, is for governance that is modest but effective in preserving the shared values at stake in a region-sized undertaking. Public values and public responsibility in the Northwest have been influential since the New Deal brought river basin development on a massive scale, together with the social promise of economic progress through public power. The burden of public values looms large in the future. As we discuss below, institutional structures adequate to these public responsibilities, and to the extraordinary complexities of the technical and economic structure, are not easy to devise.

It seems possible, nonetheless, to offer recommendations which can be constructive and concrete:

(1) Reestablishing the framework of governance is imperative *because* conflict will continue. The design of this framework must not divide political power so as to thwart leadership. Since the Bonneville Power Administration is the only institution with the expertise and historical presence capable of exercising regional leadership in the short run, it provides a natural place to begin. Yet Bonneville should not rule unchallenged.

(2) The broadened agenda of debate requires meaningful participation by those outside the traditional planning cadre of utilities, major industrial users, and Congress. Representation of state and local governments is needed, to encourage improvements in conservation efforts. Consumer representation is vital too; costs will increase--but they should increase equitably, and as slowly as possible. And environmental interests must be represented, so that economic costs can be balanced against environmental impacts. We recommend the formation of a Bonneville Advisory Council, with membership from state and local government, business and utilities, and consumer and environmental groups. The council should be charged with recommending a regional power and conservation program, scrutinizing regional forecasts, and conducting long-range studies of relevance to regional energy policy and its impacts. The Bonneville administrator should still remain a principal decision maker--but he should be required to explain publicly and in writing whenever he chooses not to follow council recommendations.

(3) The most urgent need is for regionwide conservation, not only in industrial and residential use, but in irrigated agriculture as well. Unexpectedly rapid growth in the 1970s, combined with delays in building generating facilities, may well lead to shortages of electric power in the next several years. Vigorous, coordinated, adequately financed conservation is imperative. Legislation authorizing BPA to provide technical assistance and financing for conservation is needed. Bonneville's "self-financing" capability can overcome legal barriers and bureaucratic timidity in the financing of conservation investments. We therefore believe that BPA is a natural candidate to lead regional conservation--an effort that will require the cooperation of states, local governments, and the utilities.

(4) Although there is enormous potential for conservation in the Northwest, it is probable that growth in demand will continue. With the full development of Columbia River hydropower there is no longer a federal mechanism to finance power supplies. The U.S. Senate has passed legislation to expand federal financing, in the form of "Bonneville purchase." Legal authority for BPA to contract for power plants' productive capability provides a means for organizing regional investment in generating facilities. In the coming decade BPA purchase is likely to be the lowest cost financing approach. But we do not favor permanent expansion of the federal role for this essentially regional mission. Instead, purchase can be delimited by a "sunset clause" that would insure full congressional review late in the 1980s.

(5) State and local governments must strengthen their capabilities, both executive and legislative: energy will require sustained attention well into the future. A regional power development program like that enacted by the Senate will inevitably reduce local utility control over costs of power, decisions to invest, and generally increase regional interdependence. Local and state authorities must as a result be prepared to scrutinize regional plans and to participate in regional decision making. Moreover, conservation, regulatory reform, siting, environmental protection, and a host of other policy issues demand sophisticated analysis and capable administration. The record in the Northwest has been mixed.

(6) The allocation of federal hydropower should be made more flexible and brought under the joint control of the Bonneville administrator and the new Bonneville Advisory Council. The historical claim of the public power agencies to federal hydro is beyond dispute. But a permanent subsidy to publicly owned utilities increasingly conflicts with the goals of social development and multiple use of the Columbia River that underlay New Deal investment in the Columbia Valley. In the future, the federal hydro resource can provide an incentive for public utilities to innovate in the tradition of public service that has graced their history. We advocate periodic reexamination of the ways in which publicly owned utilities advance the need for conservation and enhanced multiple use of the Columbia. Where these ends are served well by a utility, its allocation should be preserved, an economic reward to effective public service. Where a utility does not serve the long-term interests of the region, its allocation should be phased down gradually. This would provide a strong but not abrupt signal that regional objectives need attention.

(7) The technological flexibility and low cost of federal hydropower means there will be continuing economic pressure to use hydro for peaking--generation of power to meet the variable component of demand. The *fluctuations* in reservoir levels and river flow rates will be harmful to wildlife, inconvenient to recreational and navigational users, and potentially fatal to the salmon fishery of the Columbia and Snake Rivers. It is necessary to design and implement regional load-management policies, policies motivated by historic considerations of multiple use, rather than power economics alone. For instance, irrigation pumping should be curtailed during peak hours. Even with such policies in place, there will be considerable uncertainty remaining in the impacts of hydro operations. Accordingly, we recommend a federally chartered scientific task force, charged with three tasks: long-range studies of the environmental impacts of power development; technical oversight of river management; and developing alternative ways to reduce the peaking burden on the Columbia and its major tributaries.

(8) The dramatic increases in thermal-plant costs highlight the need for careful investment and close

management. As it is, BPA has estimated that the capital costs of the regional power supply will increase by a factor of six between 1979 and 1995.[24] We believe it sensible to compare each major Northwest power project against all other power plants begun in the same year elsewhere in the United States. If a Northwest project exceeds the average unit cost of others in its year class, rigorous public review should be undertaken immediately, so that costs can be contained and uneconomic projects terminated.

(9) The largest, most controversial industrial consumers of electric power are the direct-service industries (DSIs)--mostly aluminum reduction plants. The DSIs have a legitimate role in the Pacific Northwest. But efforts should be made to bring their power purchases into the marketplace, so that direct-service industries can compete on an economic basis comparable to other consumers of wholesale power. The DSIs should be neither whipping boy nor favored customer. Historically they have provided an economic service to the region by purchasing reserves, but it seems increasingly probable that portions of these reserves can be sold to utilities in the future. This is one arena in which the marketplace can choose more wisely than government.

These recommendations are built on the analysis conducted in the chapters following. Chapters II through V trace the historical influences shaping the task of institutional redesign. Chapter II discusses the hydroelectric era, stretching from the New Deal to the mid-1960s. The crucial role of the policy of public preference has its origins in the federal programs which began to transform the Pacific Northwest during the Great Depression. Chapter III discusses the Hydro-Thermal Power Program, and the technological and economic framework on which the era beyond full development of hydropower was to be built. This effort did not succeed, in part because growth in electric power came under scrutiny by new interests outside the utility industry. Those concerns, and their increasingly undeniable claims on public authority, are the subject of Chapter IV. Chapter V summarizes the several attempts at institutional redesign undertaken when it became clear that the Hydro-Thermal Program would not proceed. Since September 1977 these efforts have sought relief in congressional legislation. This evolutionary path provides the groundwork for an analysis of institutional restructuring, undertaken in Chapter VI. We develop a theory of institutional design in Chapter VII, and use it to evaluate the alternatives which appear to be available today.[25]

A decade ago the addition of thermal power to the Pacific Northwest's hydro system seemed a straightforward technological and managerial issue. Today that program is beset with unanticipated problems: fiscal, environmental, widespread mistrust of technological solutions and their technocratic managers, and a resurgence of the populism that gave rise to the public power movement and its keystone, the preference policy in the allocation of federal hydropower. Just as the national energy crisis, which strikes at the roots of an affluent industrial civilization, is a harbinger of a more modest, less predictable future, the region's electricity debate signals an era whose hallmarks are skepticism about technological growth, a groping for the uncertain potential of conservation, and a rationing of scarcity rather than distribution of abundance. We are entering a world where political turbulence and conflict arise from the sense of threatened well-being. But the choices faced by the Pacific Northwest also present an opportunity, an opportunity to realize the potential of technological civilization as an instrument of social harmony--the very potential that has brought the region to this turning point.

CHAPTER II
The Hydroelectric Heritage

Continuity with the past is not a duty;
it is merely a necessity.[1]

--Oliver Wendell Holmes

In the Pacific Northwest electric energy and its economics have been dominated by hydropower. About 80 percent of the region's electricity is supplied by falling water, and over 40 percent of the hydro potential of the nation is located in the Columbia River and its tributaries. Most of the major dams in the region have been constructed by the federal government, in an ambitious multi-purpose program. Electricity, irrigation, flood control, navigation, recreation, and other benefits have all resulted from development of the river. This large-scale program has contributed to unusually low electric power costs. But the diverse interests entwined in these multiple purposes have led to competition for the resources and potentials of the Columbia River basin.

The largest stake in this competition is the hydropower itself. It is a resource of extraordinarily low cost and unusually high desirability for technical reasons; its size and value is unmatched in the nation's electric power supply. Federal hydropower--the bulk of the Columbia system's output--is allocated under a historical policy favoring publicly owned electric utilities. Under existing arrangements the "preference policy" will continue to shift the economic benefits of the hydroelectric heritage to customers served by public power agencies. Public power serves a majority of the residents of Washington, while Montana, Oregon, and Idaho are largely served by privately owned power companies. The unequal distribution of federal hydro creates economic--hence political--conflicts. The evolution of the regional power system under the influence of the preference policy has built up a complex texture of cooperation and competition. It is a complexity which will shape, perhaps decisively, the future of the Pacific Northwest. The immediate concern is for the valuable federal hydroelectricity. But the lasting significance of the debate lies with the oldest question of regional development--how to govern the disposition of and planning for regionally pooled electric power.

In this chapter we sketch the hydro system and its principal actors, and outline the path of institutional evolution which has accompanied its growth. The demands of technology and economics compellingly urge regional cooperation in power generation; national politics, both liberal and conservative, have, for different reasons, deflected and slowed the integration of power management at regional scale. These tensions, altered somewhat by the emergence of distinctively regional issues, are the political legacy of the hydroelectric heritage.

The Pacific Northwest, for our purposes, includes the states of Idaho, Oregon, and Washington, and that part of the state of Montana lying west of the continental divide--as shown in Figure 1. Apart from small, sparsely populated corners of Nevada, Utah, and Wyoming, the major area excluded from this definition is the upper basin of the Columbia River lying in British Columbia, the westernmost province of Canada. The drainage of the upper Columbia figures importantly in the evolution of hydropower in the Northwest. Nevertheless, we have excluded Canadian and provincial interests from this analysis for the sake of simplicity: there is a complex international history which is separate from the story told here.[2] The electric power service area of regional concern is the West Group Area of the Northwest Power Pool (Fig. 2), served by the federal transmission grid of the Bonneville Power Administration. The areas excluded from the West Group are principally those served by Idaho Power Company and Utah Power and Light Company--investor owned utilities with a tradition of independence from federal government.

THE HYDRO SYSTEM

The Columbia is unique among the major rivers of the world: its great hydroelectric capabilities are located close to a large industrial and urban market for electricity.[3] Two principal supplies of electric power serve this region of 300,000 square miles, with its

Figure 1. *West Group Area of the Northwest Power Pool.*

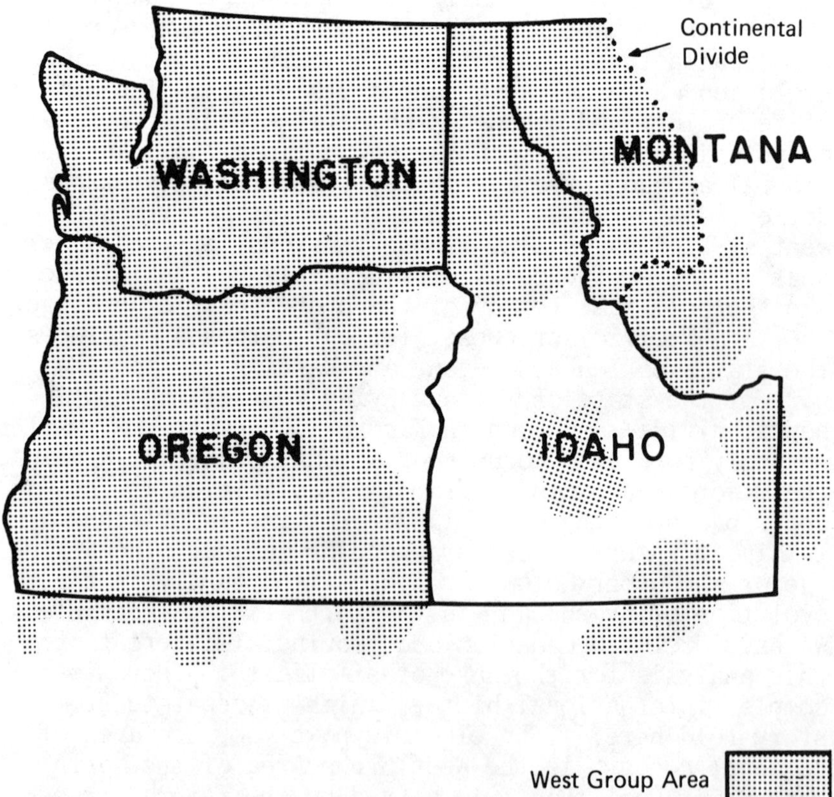

Source: *Pacific Northwest Power Outlook Through 1985-86.* (Portland: BPA, 1975) p. 1.

population of roughly seven million.[4] The first is hydroelectric power. The largest resource, accounting for more than half the total power generated, is the Federal Columbia River Power System comprising 30 hydropower projects and a regional transmission system which transports bulk power throughout the Pacific Northwest. The federal system represented, in 1978, an investment of $5.7 billion.[5] The 16 thousand megawatt (kMW) generating capacity of the federal Columbia system is augmented by 10 kMW of hydro capacity in projects developed by public power agencies and investor-owned utilities.[6] The second major source of electricity, smaller at present, is thermal power, produced in fossil and nuclear fueled plants, with a capacity of over 6 kMW.

The average power demand for 1977-78, as projected by the Pacific Northwest Utilities Conference Committee, was 16.3 kMW, of which 12 kMW, or 74 percent, was to be supplied by hydroelectric generation.[7] This average demand amounts to 142 billion kilowatt-hours a year; in fiscal year 1978 the major utilities of the region surpassed this output handily, producing 160 billion kilowatt-hours.[8] Slightly more than half the electricity was produced at federal hydro projects and associated thermal power plants; about one-fourth each was produced by investor-owned utilities (IOUs) and public agencies. (See Fig. 2.) This power served end uses as shown in Figure 3: roughly half the electric energy produced in the Northwest is used for industrial purposes, a third for residential consumers, and an eighth for commercial uses.

Delivery of electric power to final users is made by three groups of sellers. (See Fig. 4 and Table 1.) The first is the federal government: roughly one-sixth of the region's output of electricity is sold directly to seventeen major manufacturing firms--known as direct-service industrial customers or DSIs--by Bonneville on behalf of the U.S. government. These sales are made on the basis of twenty-year contracts for federally produced hydropower. Most of these contracts expire in the 1980s. One-fourth of the power is sold by 116 publicly owned utilities: rural electric cooperatives, public utility districts (PUDs) organized at the county level, and municipalities. These public agencies have

Figure 2. *Power Supply in the Pacific Northwest, Fiscal Year 1977.*

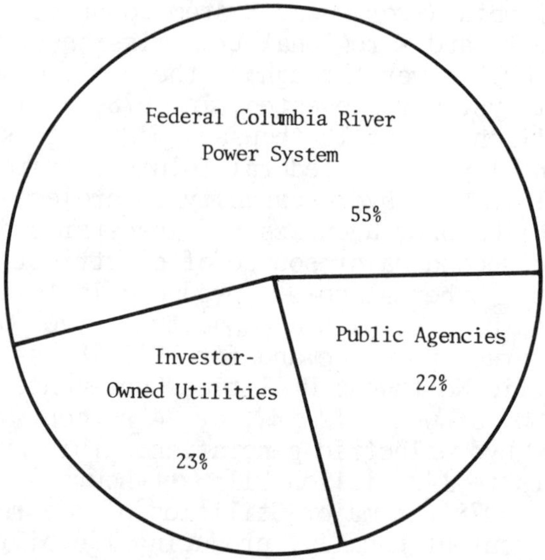

Source: Bonneville Power Administration, *1976 Statistical Summary* (Portland: BPA, 1977) p. 12.

privileged access to federal hydropower through the public preference policy. Six public agencies--serving the cities of Seattle, Tacoma, and Eugene, and Snohomish, Cowlitz and Clark Counties--provide power to a major portion of the region's customers. Of these, only the Eugene Water and Electric Board of Oregon serves customers outside the state of Washington. Half of the region's remaining electric power is sold by seven investor-owned utilities (IOUs), privately owned and vertically integrated firms. The remainder, usually less than 10 percent, is sold to users outside the region and transmitted over the federally owned power grid.

Table 2 shows that the distribution of retail sales of electric power is uneven. This is politically important in an era in which retail power costs for customers of investor-owned utilities tend to be higher than those in public agency service areas, as shown in Table 4. The geographical distribution of growth is also uneven. Some utility service areas--including the Seattle and

Figure 3. *Use of Electric Energy in the Pacific Northwest.*

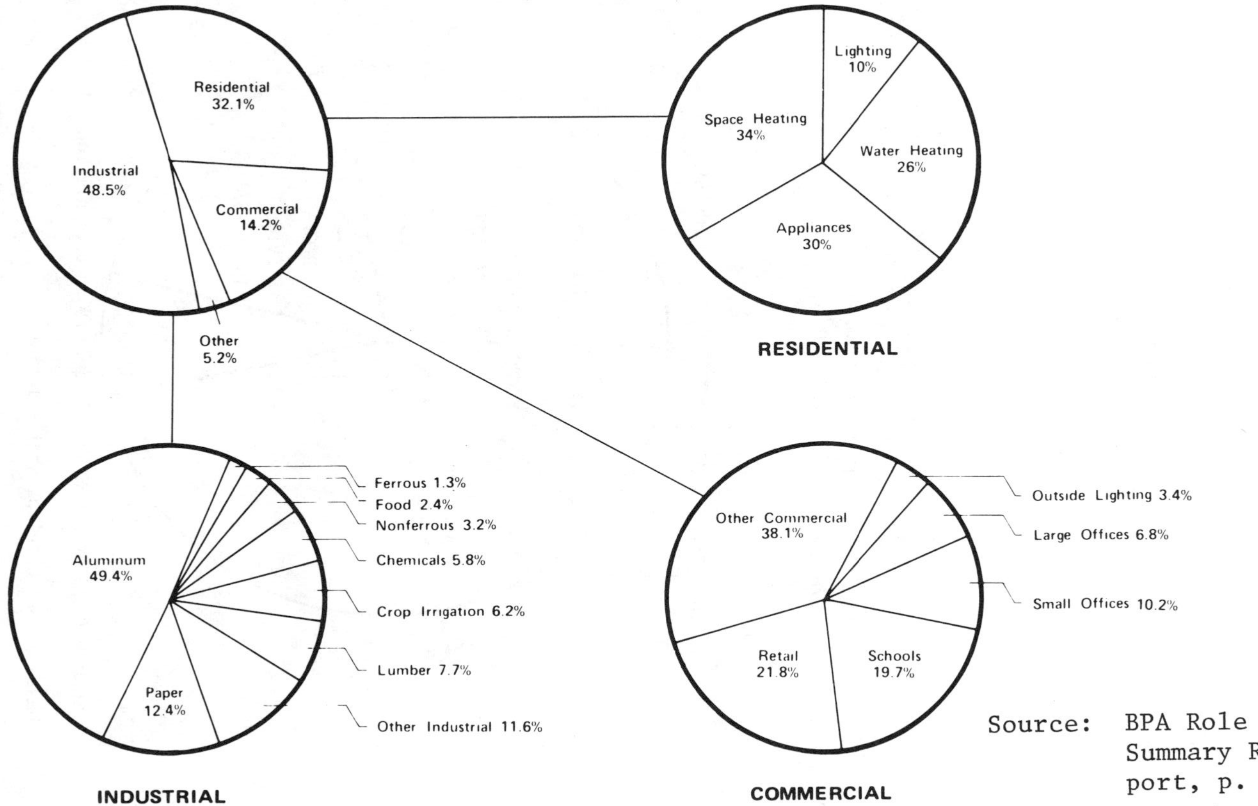

Source: BPA Role EIS, Summary Report, p. 7.

Portland suburbs and those agricultural districts where irrigation is expanding--are experiencing rapid growth of demand and expecting more. Other areas are growing slowly or holding steady. Their interest in regional investment in power resources tends accordingly to focus more on the costs than the benefits of expanding electrical supply.

Figure 4. *Final Sales and Exports, Together with Generation Sources for the West Group Area, Fiscal Year 1976 (based on 157.3 billion kwh total production).*

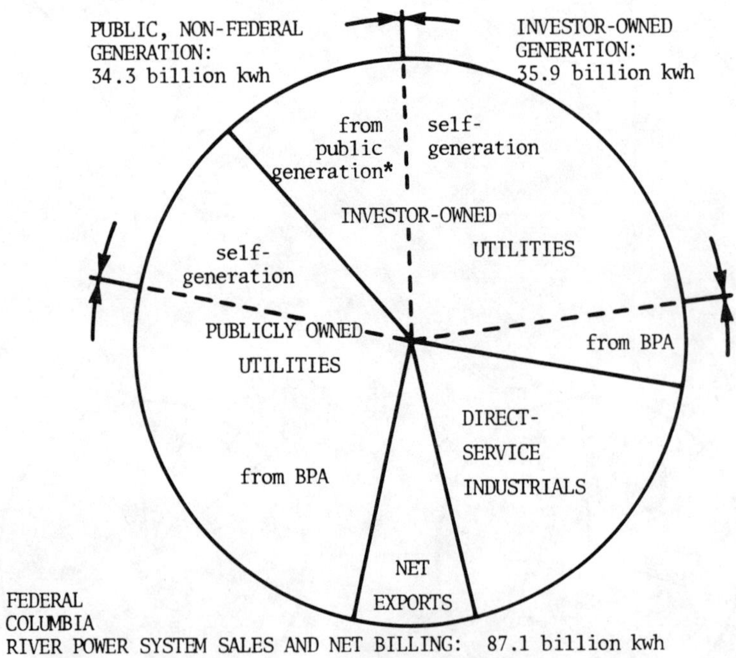

* based on BPA wheeling statistics only

Sources: BPA, *1976 Statistical Summary* (Portland: BPA, 1977) pp. 12, 14-15; Branch of Power Supply, Bonneville Power Administration, 1978.

Production: Cost and Structure

The Northwest's power supply contrasts strikingly with the national picture in industrial organization, market structure, and cost.[9] Most important, the cost of power is very low. Wholesale power generated at federal dams costs as little as 0.12 cents per kilowatt-hour--1.2 mills/kwh in technical parlance. This figure should be compared to costs of 12-14 mills/kwh for bulk

Table 1. *Wholesale Customers of BPA, 1975.*

	CUSTOMER CLASS	NUMBER	LOCATION	SELF-GENERATION	CUSTOMERS PER UTILITY (range)
Public Utility	Municipal	35 cities 2 irrigation districts 1 joint operating agency	Wash. 15 Ore. 10 Idaho 8 Calif. 5	Seattle 77.9% Tacoma 62.5% Centralia 51.1% Eugene 24.8% Idaho Falls 11.2% Bonners Ferry 39.2%	60 (Minidoka, Idaho) to 262,034 (Seattle, Wash.)
Public Utility	Public and People's Utility District	26	Wash. 22 Ore. 4	Pend Oreille 97.7% Chelan 74.3% Grant 29.3% Douglas 2.3%	108 (Alder Mutual Light, Wash.) to 14,498 (Inland Power and Light Wash.)
Public Utility	Cooperative	50	Wash. 1 Ore. 10 Idaho 15 Mont. 5 Calif. 1 Nev. 1 Utah 1 Wyo. 1	Lower Valley 4%	1,293 (Kittitas County PUD, Wash.) to 110,281 (Snohomish County PUD, Wash.)
Investor-Owned Utility		11	Wash. 2 Ore. 2 Idaho 1 Mont. 1 Calif. 4 Utah 1	All IOUs	46,545 (California Pacific Utilities Co.) to 597,231 (Pacific Power and Light Co., Ore.)
					DEMAND PER CUSTOMER (range)
Direct Service Industrial		17 (with 22 plants)	Wash. 12 Ore. 8 Mont. 2		5.5 MW (ITT Rayonier, Port Angeles, Wash.) to 444.8 MW (Kaiser Reduction, Spokane, Wash.)
Government Agency		10	Wash. 6 Lower Columbia Region 1 Mid-Pacific Region 1 State of Calif. 1 British Columbia 1		

Source: BPA Role Appendix C, pp. I-13 to I-21, and Attachment B, Table 1.

Table 2. *Electricity Market Shares and Statewide Retail Prices for Residential Customers.*

State	Electricity Market Share (1976)		Statewide Average Residential Price (1975)
	investor-owned	public agencies	
Idaho	86.1%	13.9%	16.4 mills/kwh
Oregon	62.9%	37.1%	16.5
Washington	23.4%	76.6%	12.3

Sources: Market shares, BPA Branch of Power Resources, January 24, 1978, quoted in U.S. Department of Energy, Office of the Regional Representative, "Regional Electric Power Planning Issues in the Pacific Northwest," Seattle, June 1978, p 6; prices from James McFadden, Energy Supply and Environmental Impacts, Study Module III-A, Northwest Energy Policy Project (Springfield, Va.: National Technical Information Service, 1977) pp. V-13 to V-15.

power from contemporary thermal plants.[10] (See Table 4.) Thermal plant costs have mounted rapidly. Plants now under construction, scheduled for operation in the mid-1980s, will have power costs ranging around 40 mills/kwh. The extremely low cost of hydropower reflects a number of factors: the bounty of the potential energy in the waters of the Columbia River; the cost-sharing inherent in multiple-purpose development, allowing uses like navigation and flood control to dilute the sums charged against power production; and the low interest rates, now ranging from 2.5 to 6.6 percent, charged by the U.S. Treasury to finance federal projects.[11]

Under the public preference policy, the bulk of BPA-marketed power is sold to public agencies, as shown in Table 4. Investor-owned utilities, which may only purchase energy surplus to the needs of public-agency customers, now account for about 10 percent of BPA sales--down from over 20 percent in 1972.[12] Thus, the preference policy, by curtailing the power available to the IOUs, has forced these companies to develop more expensive thermal plants. (With further growth in demand, public systems are turning to thermal power generation in substantial quantity as well.)

The uneven distribution of low-cost federal hydro has substantial impact on retail power prices. Seattle customers, able to draw upon hydropower from Bonneville

Table 3. *Selected Typical Monthly Electric Bills (June 1978).*

Utility	Cost of 1,000 kilowatt-hours (residential rate)
Farmers Electric Cooperative, Idaho	$ 7.08
Seattle City Light, Washington	10.39
Clark County PUD, Washington (suburban Portland)	11.10
Snohomish PUD, Washington	11.40
Tacoma City Light, Washington	12.80
* Washington Water Power, Washington State (eastern)	14.73
Eugene Water & Electric Board, Oregon	16.57
* Idaho Power Company, Idaho	21.85
* Portland General Electric & Pacific Power & Light, Portland, Oregon	27.00 to 27.24
Prairie Cooperative, Idaho	32.40

* Investor-owned utility; others are public agencies.

Source: Bonneville Power Administration, proposed 1979 wholesale rate increase, draft environmental impact statement DOE/EIS-0031-D, August 1978, Appendix B.

and the city's own dams, pay half what residents of Portland pay to their investor-owned electric companies. But even customers of hydro-poor IOUs pay less for elec-power in the Northwest than in other parts of the nation. (See Fig. 5.) A recent estimate placed typical residential power bills in Tacoma, Washington, at one-fifth that charged residents of New York City.[13] It is useful to bear in mind, however, that wholesale power cost constitutes only part of the retail price paid by the final user; differences in distribution cost, managerial efficiency, and local taxes produce a wide variation. In the Northwest, residential rates vary by a factor of

Table 4. *BPA Wholesale Sales, Fiscal Year 1976.*

	FIRM POWER (Millions of kilowatt-hours)	PRICE+ (mills/kwh)	NON-FIRM POWER (Millions of kilowatt-hours)	PRICE (mills/kwh)
Pacific Northwest				
Public Utilities	30,701	3.76	116	3.22
Investor-Owned Utilities	2,656	6.34	5,005	3.45
Direct Service Industrial Customers				
Aluminum	22,288	3.14	396	2.77
Other	2,315*	3.27	63	3.54
Federal and State Agencies	635	3.68	2	3.89
Outside Pacific Northwest				
Public Utilities			3,515	3.72
Investor-Owned Utilities			8,294	3.36
Federal and State Agencies			1,485	3.29
TOTALS	58,595		18,876	

* Industrial firm energy (see text).

\+ Actual prices paid for delivered power from BPA.

Source: BPA, Fiscal Year 1976 Generation and Sales Statistics (Portland: BPA, 1976) pp. 3-4.

Figure 5.

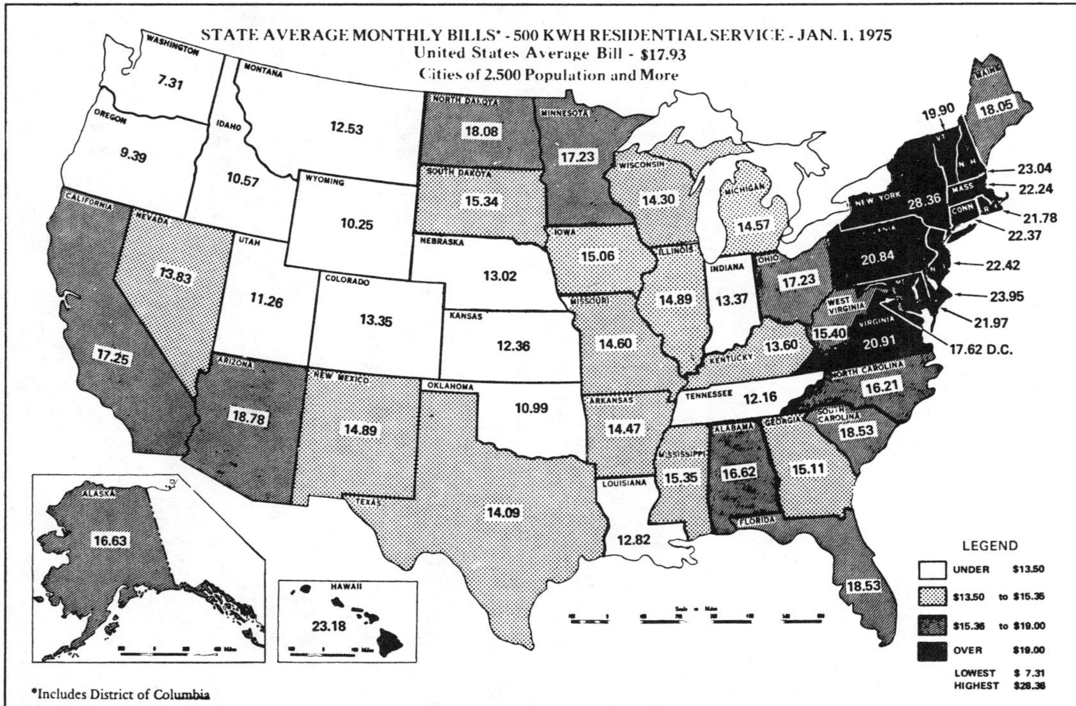

Source: Federal Power Commission, *Typical Electrical Bills*, 1975.

four, as indicated in Table 3. It is interesting that both the lowest and highest rates are charged by public agencies. What is politically significant, however, is that large public agencies, such as those serving Seattle and suburban Clark County, Washington, charge rates strikingly lower than Portland's.

Organization and Integration

A second essential feature of the Northwest power supply system is its regional and industrial organization, heavily influenced by the federal presence, BPA in particular. Although utilities in many parts of the United States have formed interconnected power pools on a regional basis, the degree of integration among the major electricity producers and consumers in the Northwest is unusual. The operational interdependence which links together the components of the regional power system is illustrated in Figure 6 by the power sales in Washington State among public agencies, IOUs and the federal generating system.

At the center is Bonneville (Fig. 7). BPA transmits power throughout the Northwest over 12,500 miles of transmission lines, providing 80 percent of the region's transmission capability for bulk power.[14] Through its direct control of most of the interconnecting links among utilities, BPA functions as a regional distributor of bulk power generated at both federal and non-federal facilities.

Bonneville's revenues for the power it supplies and the transmission services it provides are based upon the operating and capital costs of the federal Columbia system. Under its legislative mandate, BPA must collect revenues sufficient to repay the U.S. Treasury for current operating costs and for investments in the federal electric power system, mainly dams and transmission facilities. Hydroelectric dams must be paid for over a fifty-year period, and transmission equipment, over a thirty-five year time span. "Mortgage payments" are estimated each year in a repayment study, which analyzes current requirements and projects future revenues and costs, to insure that rates will be sufficient to recover investments. The wholesale rates charged BPA customers, in turn, are set to meet repayment requirements

Figure 6. *Electric Power Generation and Sales in Washington State, 1975.*

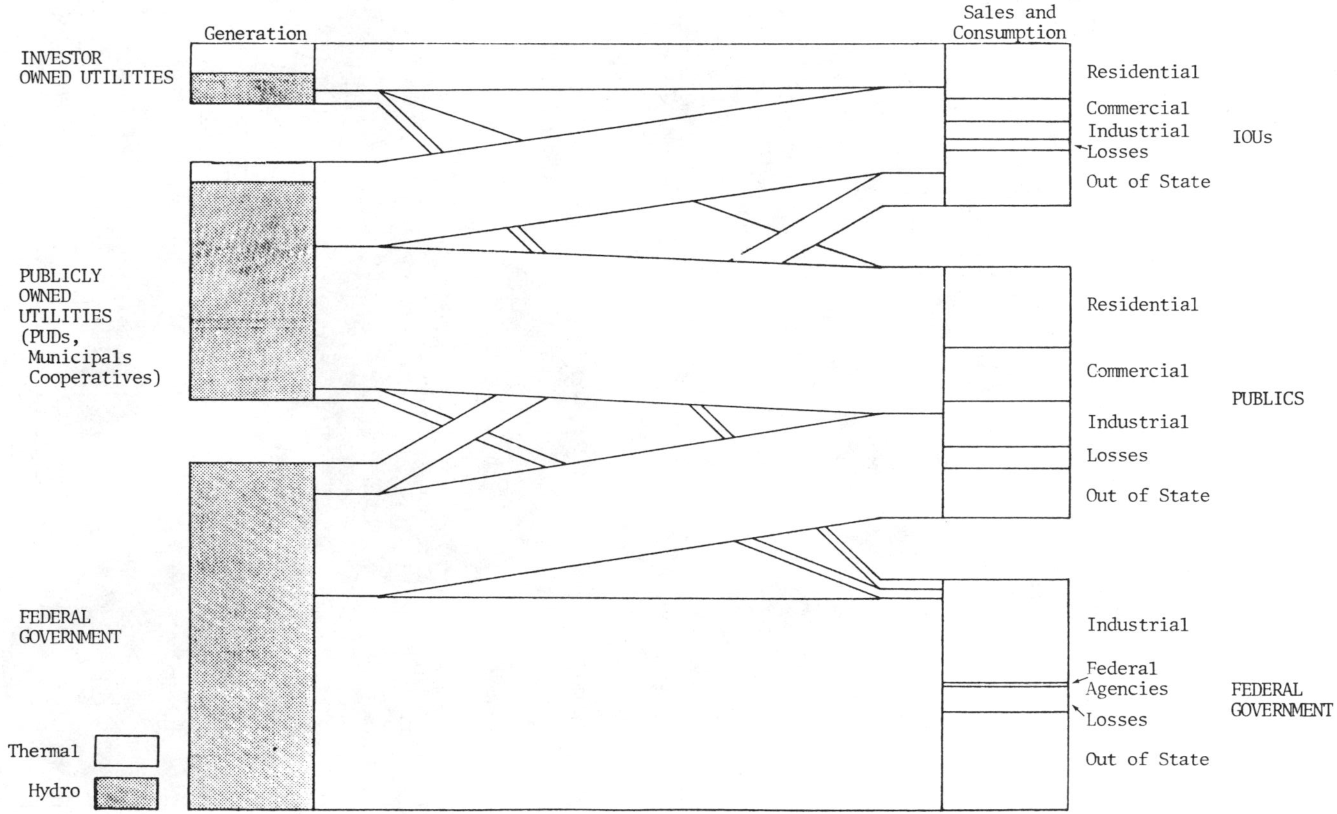

Figure 7. *Northwest Power Pool Interconnections, 1977.*

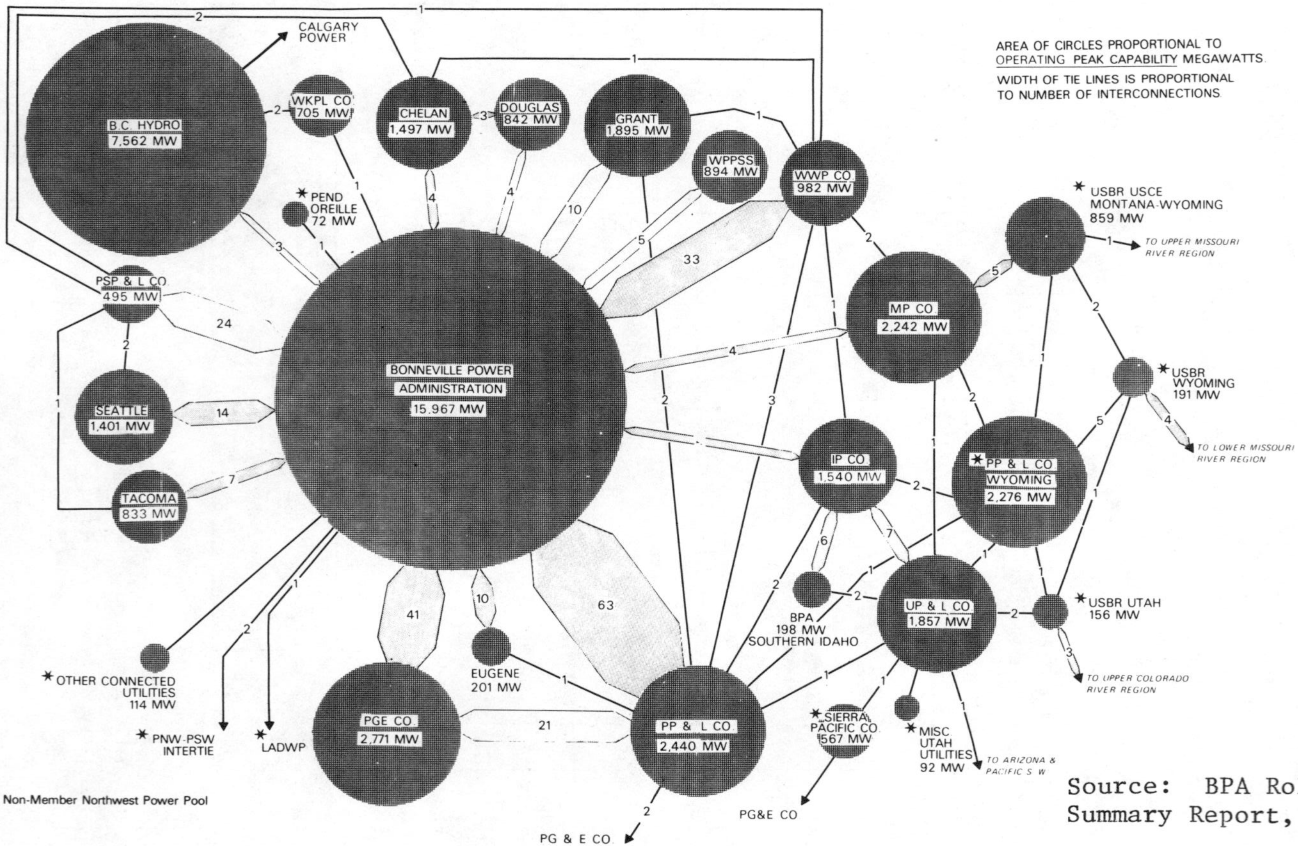

Source: BPA Role EIS, Summary Report, p. 29.

and operating costs of the federal Columbia system.[15] With the development of the thermal power net billing arrangements described in Chapter III, the revenue requirement has begun to rise rapidly, since BPA has signed contracts which in effect finance the capital costs of new nuclear power plants. As this book went to press, BPA began to implement a 90 percent increase in its wholesale power rates.

Bonneville's regional transmission grid has influenced the structure of the utility industry dramatically. Most privately owned utility companies are vertically integrated: a single corporation owns all the elements of a power supply system from generating plant to final distribution network. The IOUs of the Northwest generally adopt this form, though even the IOUs rely on the Bonneville grid for most wholesale power transfers. (As noted before, they also obtain some electric power from the federal Columbia system.)

The public power sector, with its guaranteed access to federal hydro, does not require firms of the large size demanded by the fiscal scope of vertical integration;[16] public power is accordingly highly fragmented, with 116 agencies delivering power to retail customers. In the 1940s, in an effort to assure use of the large blocks of power coming on line with the construction of federal dams, Bonneville required its public system customers to purchase 90 percent of their electricity needs from BPA in order to qualify for a full "requirements" power sales contract.[17] As a result, some utilities, such as the Snohomish County PUD, became largely or wholly dependent upon Bonneville for the supply.

Consistent with its mandate to promote rural electrification, Bonneville has assured its smallest customers--mostly rural cooperatives--of federal power sufficient to meet loads up to twenty-five megawatts. The Salmon River Coop in Idaho had a load in 1975 of only 2.7MW;[18] although its load growth over the preceding decade was a rapid 10 percent per year, the coop would not exceed its BPA ceiling until the end of the century even at that high growth rate. Without the need to invest in thermal facilities, smaller public utilities have not dealt with the complexities of planning controversial and costly facilities. With their small, relatively isolated staffs, these smaller utilities also lacked the incentive to form alliances with other utilities in the region. Their major source of information--and reassurance--has been Bonneville.

While the availability of a low-cost power supply from federal resources has fostered dependence on BPA among some publicly owned utilities, others have pursued the development of their own power supplies. When Puget Sound Power and Light Company allowed its federal permit to build a Skagit River hydro project to lapse, Seattle City Light developed the Skagit's generating potential. Now, regardless of how federal power is allocated, Seattle will maintain its own block of cheap hydro resources. Several other public utility systems, notably the Chelan, Grant, and Douglas PUDs along the mid-Columbia, together with IOUs such as Washington Water Power and Idaho Power, have developed hydro resources. These utilities are also in a favorable situation in the current regional power controversy since they are less dependent upon access to federal resources to keep their costs down.

BPA's influence in the region reaches beyond wholesale power, despite attempts by Congress and some presidents to restrain its activism. For the scope of the federal Columbia system has inevitably produced regional coordination needs, many of which have been shouldered by BPA.[19] Bonneville has traditionally carried out the analyses underlying the demand projections which form the basis of utility planning. BPA has also provided detailed guidelines for regional development in its economic base studies, industrial site analyses, county inventories, and other planning studies. Most important, BPA's technical coordination in power marketing and transmission has served increasingly to integrate the Northwest utility industry into a single regional generation system.

As the hydropower potential of the Columbia river system has been harnessed, public and private utilities have had to consider generating resources that they would have to build on their own. A natural form of cooperation would pool public and private efforts, since each form of utility organization has distinctive strengths and weaknesses. Yet the very difference in public and IOU industry structures has inhibited cooperation in investment. The regional transmission grid and BPA's regional mission have structured the cooperation that does exist through interconnection.

Connections among utilities permit collaborative operation and facilitate cooperatively scheduled investment. Thus, if the Sasquatch Public Utility District suffers a power outage, it can immediately acquire emergency power from other utilities via the regional grid. Or more routinely, knowing that a large new thermal plant is scheduled to be brought on-line by Sasquatch PUD, other utilities can contract to purchase portions of the plant output not needed immediately by Sasquatch, deferring their own construction of power plants. The ability to share equipment, either in an emergency or when there is surplus capacity, lowers operating and investment costs markedly. This is the primary motive for utilities to "make treaties with one another, called interconnection agreements."[20] Since the mid-1950s--when large-scale power development by non-federal entities began in earnest--BPA has provided "wheeling" services: the distribution of power from non-federal power projects throughout the Northwest to utilities and industrial customers.[21] The BPA grid now ties together utilities throughout the Northwest, and links them to British Columbia to the north and the American Southwest, via the Pacific Intertie.

Like other forms of entangling alliance, however, the treaties called interconnections have costs as well as benefits: "...coordination that benefits firms A and B may require new transmission lines in company D's service area."[22] The problems of sharing both the costs and the benefits derived from interconnection have traditionally required external leadership with authority sufficient to induce utilities to transcend individual self-interest. In the Northwest, BPA has played this role, encouraging power pooling both in its policies and by building and expanding the regional grid. This leadership role is regarded with some ambivalence by utilities jealous of their managerial independence, even as they reap the benefits of region-wide coordination.[23]

Occasional conflicts notwithstanding, Bonneville maintains a dominant role in the development of the regional grid. The operating voltages in BPA's transmission system range from 12.5 kV to 500 kV a.c. and 800 kV d.c. BPA is currently replacing its existing low voltage lines with a massive 500 kV grid, a project that involves approximately 1,000 miles of high-voltage transmission lines over the next ten years and up to 2,000 miles more in the rest of the century. These improvements are designed to expand BPA's ability to meet its contractual obligations to deliver and transmit power, while maintaining the reliability of the overall network and improving the efficiency of the grid.[24]

Direct-Service Industrial Customers

The third distinguishing characteristic of the regional power supply system is the role played by the DSIs in power management. Drawn by the Northwest's abundant, low-cost electricity Alcoa signed the first DSI contract with BPA at the end of 1939. Since then the industry has located 30 percent of the nation's aluminum reduction capacity in the Pacific Northwest (see Fig. 8 and Table 5.) The aluminum refiners account for 90 percent of BPA's sales to DSIs--about a third of the power marketed by Bonneville in a typical year.[25]

The region has derived important benefits from the presence of the twenty-two DSI plants. The plants provide about 15,000 jobs, with an annual payroll of $280 million in 1975; the DSIs also paid property taxes to state and local governments of $11.5 million that year.[26] Of equal consequence, the technical characteristics of their operation allow for more efficient use of generating plants, scheduling of additional capacity, and management of generating reserves.

With some advance warning, aluminum refining chambers, or potlines, can be shut down for hours or weeks at a time.[27] Shutdowns carry varying economic penalties to the manufacturer. Bonneville sells power to the DSIs, part of it on an interruptible basis, at prices substantially lower than those charged for firm, or guaranteed, power. The lower price cushions the economic losses of having to shut down potlines when power is interrupted. Interruptible power can be diverted when needed by BPA, to serve customers, such as residential and commercial users, who would be greatly inconvenienced by disruption of supply. Thus the DSIs

Figure 8. LOCATION OF BPA DIRECT SERVICE INDUSTRIAL CUSTOMERS

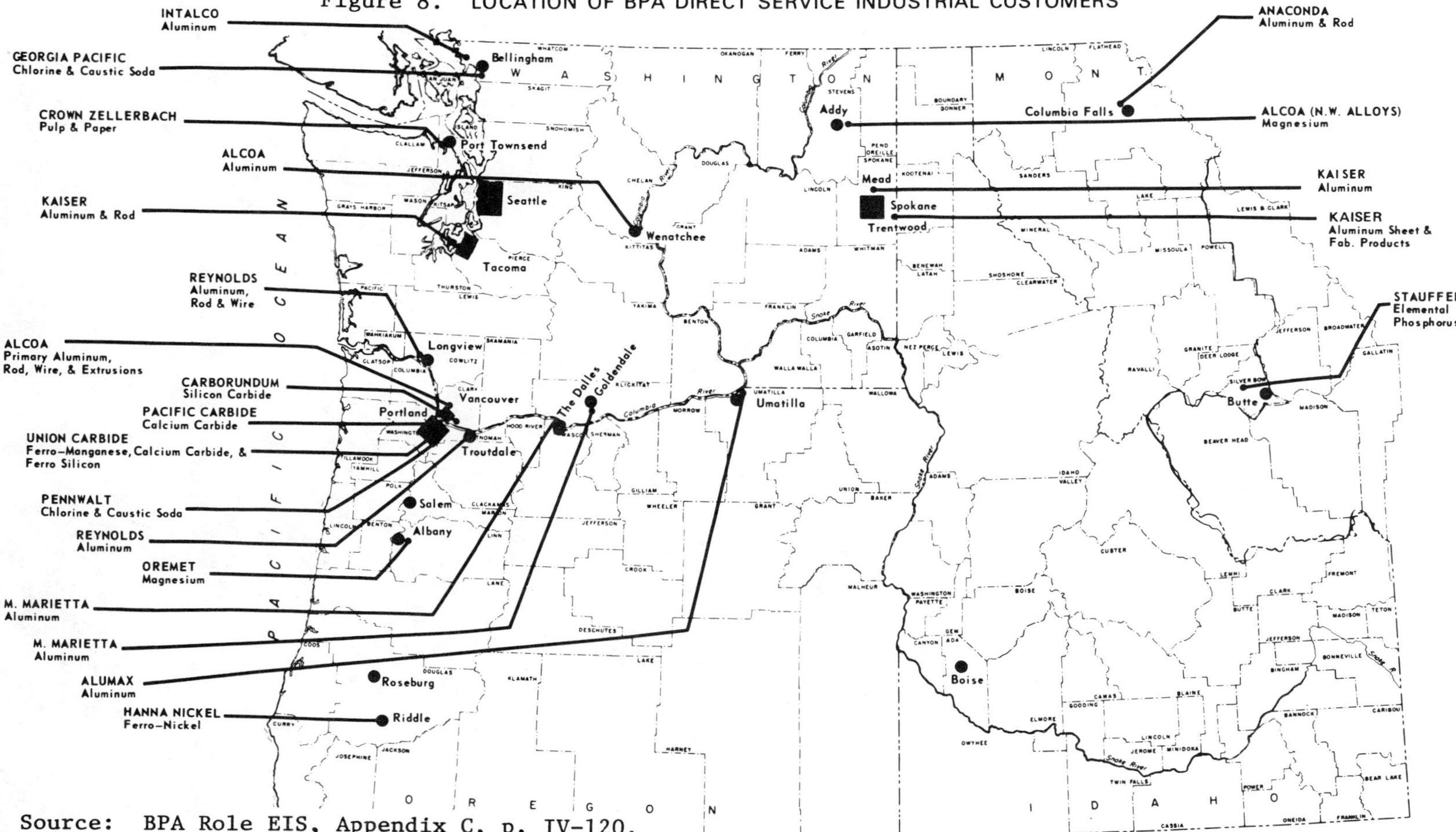

Source: BPA Role EIS, Appendix C, p. IV-120.

make use of part of the system reserves needed to hedge against low water from drought, unanticipated facility breakdowns, and unexpected surges in customer demand. Rarely has more than half of the DSI load been interrupted for any significant duration--although during the two major droughts of the 1970s substantial interruptions were required.[28]

Table 5. *Direct-Service Industrial Customers of BPA (1976).*

CORPORATION	LOCATION	PRODUCTS	MAJOR USES OF PRODUCTS
Aluminum			
Alcoa	Vancouver, Wash.	Primary Aluminum Nail, Wire and Extrusions	Construction, Transportation Equipment, Packaging, Electronic Components
Alcoa	Wenatchee, Wash.	Primary Aluminum	" " " " " "
Alumax	Umatilla, Wash.	Primary Aluminum	" " " " " "
Anaconda	Columbia Falls, Mont.	Primary Aluminum	" " " " " "
Intalco	Bellingham, Wash.	Primary Aluminum	" " " " " "
Kaiser	Mead, Wash.	Primary Aluminum	" " " " " "
Kaiser	Trentwood, Wash.	Aluminum Sheet and Fabricated Products	" " " " " "
Kaiser	Tacoma, Wash.	Primary Aluminum and Rod	" " " " " "
Martin Marietta	The Dalles, Ore.	Primary Aluminum	" " " " " "
Martin Marietta	Goldendale, Wash.	Primary Aluminum	" " " " " "
Reynolds	Longview, Wash.	Primary Aluminum Rod and Wire	" " " " " "
Reynolds	Troutdale, Ore.	Primary Aluminum	" " " " " "
Other			
Alcoa (N.W. Alloys)	Addy, Wash.	Magnesium	Construction, Alloying of Aluminum
Corborundum	Vancouver, Wash.	Silicon, Carbide	Industrial Abrasive
Crown Zellerbach	Port Townsend, Wash.	Pulp and Paper	Heavy Kraft Paper and Liner-Board
Georgia-Pacific	Bellingham, Wash.	Chlorine, Caustic, and Soda	Bleaching of pulp and paper
Hanna Nickel	Riddle, Ore.	Ferro nickel	Stainless Steel
Oremet	Albany, Ore.	Magnesium, Titanium	Aerospace industry, paint, Valves
Pacific Carbide	Portland, Ore.	Calcium Carbide	Acetylene production
Pennwalt	Portland, Ore.	Chlorine, Caustic Soda	Bleaching of pulp and paper
Stauffer	Silver Bow, Mont.	Elemental Phosphorus	Fertilizer
Union Carbide	Portland, Ore.	Ferro manganese, Calcium Carbide and Ferro Silicon	Alloying of steel, Acetylene production

Source: BPA Role EIS, Appendix C, p. IV-121.

Most of the power sold the DSIs is marketed on the IF-1 rate schedule introduced by Bonneville at the end of 1974. IF-1 power is priced at a single rate, but the power is divided into four equal quartiles. Each quartile is interruptible, but for varying causes. Bonneville may unilaterally interrupt one quartile, "for any reason, at any time, for any period." Other quartiles may be shut off to compensate for unanticipated power plant failures, to provide backup for the federal hydro system, and to provide regional reserves; this power is called "industrial firm." It is firm in the sense that, when interruptions take place, BPA provides credits to DSIs whose power is curtailed; thus, while the power may be cut off in whole or part, the economic loss from interruption is compensated partially by the credit provision.[29]

This arrangement presents the apparent anomaly of the industrial customers' being charged a lower rate because their power is not firm--but receiving credits upon interruption because it *is* firm, in the economic sense defined. The actual effects of the IF-1 schedule are difficult to analyze, however, because DSIs are reluctant to disclose their costs of operation. Moreover, the differences between industrial customers probably means that some DSIs benefit substantially from IF-1 while others do not. It is relevant in this regard that IF-1 was chosen voluntarily by a majority of the DSIs when it was offered as a replacement for the MF-1 schedule then in effect.

The direct-service industries have also provided economies of scale, whose benefits are shared by other users in the region. A 1971 estimate of the costs to BPA of serving the DSIs concluded that the cost of service was about 75 percent of the revenues received. BPA's revenues more than paid for generation, transmission, and services provided the industrial customer.[30]

Yet the DSIs have become controversial in recent years as the cost of adding new supplies of electricity has risen sharply. In the past, technical progress gradually lowered the cost of generating electric energy. As additional facilities were built, the average cost of generating a kilowatt-hour of electricity declined. Therefore, large customers moving into the

Northwest brought lower electricity prices as well as jobs and other economic benefits. Since 1970, however, this trend has sharply reversed. The costs of adding nuclear and coal-fired power stations have risen dramatically, and the Northwest has now realized the available hydro potential of the Columbia River. As the costs of expanding power supply have begun to rise, the DSIs have become a tempting target for criticism. Their traditional access to low-cost power seems to be a privilege no longer justifiable, especially to critics for whom the DSI role in providing regional reserves is so much technical arcana.[31]

The DSIs have substantial economic interests in continued access to large blocks of electric power, and they appear willing to pay prices significantly higher than historical hydropower rates. Each pound of aluminum produced in a primary reduction facility in the Northwest requires between seven and nine kilowatt-hours of electric power. In fiscal year 1976, power delivered to Northwest DSIs cost, on average, 3.2 mills/kwh, as shown in Table 4. Electricity thus constituted about 6 percent of the prevailing market price for aluminum ingots of 42 cents per pound.[32] The cost of DSI power has varied from year to year, depending upon the extent and duration of interruptions of federal hydropower; during periods of curtailment DSIs sometimes purchase replacement energy from non-federal sources, typically at prices of 8-12 mills/kwh.[33] Plants in other regions of the country pay 10-12 mills/kwh all the time, however, so Northwest DSIs still enjoy a substantial advantage. This advantage is partly offset by transportation costs. Two-thirds of the ingot produced in the region is shipped to the Midwest for forming and fabrication--including aluminum destined for Boeing aircraft assembled in Seattle.

The combined impact of power and transportation costs varies among DSIs. As part of the BPA Role EIS, the consulting firm of Ernst and Ernst estimated the economic differential between a hypothetical "average" Northwest aluminum plant and a competitor, also hypothetical, in the Ohio Valley. The Ernst and Ernst analysis, using incomplete data supplied by industry, concluded that existing Northwest facilities enjoyed an advantage in overall production and transportation costs of more than 10 percent.[34] It follows from this that average power prices in the range 9-12 mills/kwh could be charged the DSIs before their regional advantage would be dissipated.[35] Up to 25 percent of the market price of aluminum ingot would then represent costs of electric power. Some reservations should, however, be noted. First, the data base for the Ernst and Ernst study cannot be independently scrutinized, since it included proprietary information the DSIs were unwilling to disclose publicly, apparently for fear of losing competitive advantages. Second, the hypothetical "average" aluminum plant does not accurately portray the situation of any one of the firms operating plants in the region. Some, such as Alcoa, have plants elsewhere; over time, they could probably shift operations in response to power price changes without undue difficulty. But Intalco owns only the plant sited in the Northwest. Moreover, the degree to which existing plant and equipment has been amortized varies, though much of it has been paid for by past production of aluminum.[36]

The fact remains that DSI contracts are due to expire in the 1980s and early 1990s (Table 6). The sizable quantities of power which could be reallocated to other consumers makes the industrial customers a tempting political target. (The DSIs have responded vigorously to public expressions of concern about industrial power sales with publicity campaigns stressing the economic and environmental contributions made by aluminum.) Moreover, as Table 6 suggests, each DSI plant is located in the service area of a retail utility, obliged to provide power to any customer requesting it. The legal, technical, and economic ramifications of transferring the DSIs to retail service have not been examined publicly in detail, although as one industry spokesman noted, "Some DSIs might actually benefit from the low rates publicly-owned utilities are able to charge their industrial customers."[37] Retail connection could further exacerbate the problems of allocating low-cost hydropower, since nearly 85 percent of the DSI load is located within or adjacent to public power service areas.[38] These plants may therefore be eligible for allocations of federal hydro.

It is worth noting that significant economies are still conferred on the region as a whole by the DSIs,

Table 6. *Direct-Service Industrial Customers of BPA, Showing Contract Expiration and Utility Serving Each Plant Location.*

DSI CUSTOMER	DATE OF ORIGINAL SERVICE[a]	CONTRACT EXPIRATION DATE[a]	(MW) CONTRACT PERIOD-1975[a] (Avg. MW)	LOCATION	SURROUNDING or ADJACENT UTILITY[b]
Union Carbide Corp	7-1-42	5-11-81	21.5	Portland, Ore.	IOU
Crown Zellerbach	10-21-49	8-30-83	13.7	Port Townsend, Wash.	IOU
Georgia Pacific	9- -65	7-6-84	23.7	Bellingham, Wash.	Public
Intalco Aluminum	4- -66	10-22-84	420	Bellingham, Wash.	Public
Carborundum Corp.	4-8-49	12-31-85	26.8	Vancounver, Wash.	Public
Penn Walt Corp	6-3-47	12-31-85	45.4	Portland, Ore.	IOU
Kaiser Reduction	6-19-46	10-10-86	444.8	Spokane, Wash.	Public
Reynolds	8-19-41	12-28-86	411	Longview, Wash.	Public
Reynolds	9-28-46	12-28-86	274.4	Troutdale, Ore.	IOU
Alcoa	9- -40	6-15-87	246.6	Vancouver, Wash.	Public
Alcoa	6-16-52	6-15-87	201.9	Wenatchee, Wash.	Public
Anaconda Aluminum	2-21-55	8-9-87	335.9	Columbia Falls, Mont.	Public
Martin Marietta	11-7-57	2-13-88	186.8	The Dalles, Ore.	Public
Martin Marietta	7- -71	2-13-88	213.2	Goldendale, Wash.	Public
Stauffer Chem. Works	11-30-51	4-22-88	66.7	Silver Bow, Mont.	Public
Oregon Metallurgical	4-17-70	5-7-88	7.3	Albany, Ore.	Public
Hanna Nickel Smelting	6-10-54	6-26-90	113	Riddle, Ore.	IOU
Pacific Carbide and Alloys	12- -40	9-9-91	16	Portland, Ore.	IOU
Alumax		7-20-86	320[c]	Umatilla, Ore.	IOU
Alcoa		6-15-87	53[d]	Addy, Wash.	IOU
Kaiser Rolling	6-1-46	10-10-86	60.8	Spokane, Wash.	Public
Kaiser Reduction	1-9-47	10-10-86	149.2	Tacoma, Wash	Public

a. From BPA Role EIS, Appendix C, Attachment B, Table 1, pp. B-5, B-6.

b. From Bonneville Power Administration, Branch of Rates, "Analysis of Rate Directives," draft, February 1979, p. 13. DSIs merely adjacent to public utility areas have been labeled "public."

c. The Alumax plant has not yet been constructed.

d. Interruptible. This plant, too, is not yet on line. Tom Naguchi, interview by DLK, 2/15/78.

as the region moves beyond principal dependence on hydropower. Thermal plants can be more economically and safely operated under baseload conditions--that is, steadily around the clock. Aluminum potlines are highly automated and typically operate twenty-four hours a day. DSI loads are therefore technically desirable to power managers to help smooth the daily and hourly variations of power demand.[39] The DSIs also provide financial benefits by purchasing interruptible power used as regional reserves; this is power that has had no alternative market since it must be instantly available to back up operating power plants.

In sum, in technical operations the hydro-based Pacific Northwest power system is a regionally integrated network, coordinated under BPA leadership, and comprised of two groups of wholesale customers: industrial bulk power users, mainly in the aluminum industry, and utilities under both public and private ownership. These characteristics are the result of forty years' steady evolution, as the Columbia River was harnessed, and then turned into the centerpiece of a gigantic power supply system spanning the Northwest.

THE HYDROELECTRIC PERIOD: 1937-1964

Although the Minidoka generating project on the Snake River went into service in 1909, the harnessing of the hydropower potential of the Columbia River system was essentially a product of the quarter century from the Great Depression to the New Frontier. This enormous engineering feat was a public undertaking in the main, and the political controversies which affected its timing and influenced the design of its managing institutions reflected the shifting national debate over public responsibilities for social ends. The forces of technology and politics have tended toward regional integration, with widening inclusion of affected interests.

What began as a national government commitment to spur development has become a complex regional network of technical, business, and political actors. They confront and clash over a many-faceted, sometimes contradictory agenda of social, environmental, and economic concerns. The paradox, we shall argue in Chapter IV, is that the broadening of participation has clouded our sense of public purposes: when regional power development began, the ends it served were far clearer than the futures we think now to choose among.

Democratic Presidents and Congresses have seen the Pacific Northwest as an arena where publicly sponsored electric power could bring about region-scale economic and social progress. Republicans have been suspicious of government-owned electricity, preferring to put investment decisions in the hands of private enterprise. Growth has been a bipartisan goal, however, and the institutional arrangements for planning and operating the hydropower system have responded--if not always smoothly --to the imperatives of technology and economics.

Public Power

By the time the Bonneville Project Act was signed by President Roosevelt in 1937, public power had already won a strong beachhead in the Pacific Northwest.[40] The state of Washington was the center of the movement: publicly owned municipal utilities were firmly established in the two largest cities, Tacoma and Seattle,[41] and in 1930 the Public Utility District Act had institutionalized the processes by which communities could move rapidly and easily into public ownership of electric production and distribution. The major missing ingredient was an abundant supply of power. The Bonneville Act remedied this lack by establishing the arrangements for marketing power from Bonneville Dam, authorized by Congress in 1933, and operated by the U.S. Army Corps of Engineers.

Written in the waning days of the New Deal, the Bonneville Project Act declared strongly in favor of public ownership of power generation.[42] But, reflecting the political compromise which enabled its passage, it gave the fledgling Bonneville Power Administration authority only over the marketing of electric power, and placed BPA financing under Congressional appropriations and review. Limitations on its mission have led BPA over the years to engage in considerable bureaucratic jockeying with other federal agencies, especially the U.S. Bureau of Reclamation.[43] As the dams were built, BPA's marketing authority and service area were successively expanded until now BPA is the principal federal presence in the region's electrical supply system.

Bonneville Dam and its larger upstream cousin, Grand Coulee, were the first long steps toward the harnessing of the Columbia River. In those years of economic depression, large federal public works investments were a major element of the Roosevelt Administration's attempts to restart the stalled economy through public spending. But the decision to focus on public development of hydropower reflected two other premises of social policy as well.

First, the turbulent, scandal-plagued experience of the electric power industry during the 1920s had led to the idea that public production of electricity could serve as an economic "yardstick," providing a competitive measure of the soundness of investor-owned utilities' operations.[44] Proponents of the yardstick concept argue that government regulation of the electric power industry is inadequate. Government development and ownership of electric generation and distribution systems is also needed, to provide examples of how low costs can be kept. This applies competitive pressures to IOUs, both through the support given regulatory authorities' determination of prices and directly, through competition for new customers.[45] Franklin Roosevelt became personally committed to this view as governor of New York; he led a successful battle to create the Power Authority of New York, a public power agency established in 1931 to compete against privately owned electric companies.[46]

Economists disagree on the effectiveness of yardstick competition. The basic complication is that, as things have turned out, public and private enterprise operate in different environments and do different things. Public power agencies typically do not pay taxes, and they enjoy lower interest rates from lenders confident of government's creditworthiness. In addition, public power development has often focused on multiple-purpose hydroelectric projects aimed at broad-scale economic development. Such differences in operating environments and objectives make it difficult to carry out the yardstick comparison unambiguously. One exhaustive study concludes that "utility managements, when exposed to government competition, have lowered prices"; a textbook on utility economics finds, however, that "the yardstick comparison has little economic validity.[47]

The second premise which supported hydro development is the idea that low-cost electric power, intelligently developed on a regional basis, would serve as a transitional resource, enabling undeveloped and rural regions of the nation to develop manufacturing industry. In this way, they could move beyond the economic instabilities of agriculture and resource exploitation characteristic of "less developed" areas and enter the mainstream of industrial America. Hydropower provided a crucial comparative advantage to the Pacific Northwest, since it could be produced at significantly lower cost than elsewhere in the nation. The aluminum industry, drawn to the region by abundant, low-cost power, thus symbolized the industrial takeoff of the Northwest, and the shift away from the boom-bust cycles of lumbering, agriculture, and fisheries.[48]

Carrying the freight of these social hopes and ideological commitments, the Bonneville Project Act[49] made four policy declarations which continue today to resound in the political arena. First, the Bonneville administrator was to "encourage the widest possible use of all electric energy that can be generated and marketed." (The original intent was to provide public power competition to investor-owned utilities as new areas were electrified. But in an era when conservation has become a byword, widest use may no longer be wisest use.) Second, Bonneville Dam and, later, other federal projects were "to be operated for the benefit of the general public, and particularly of domestic and rural customers." This identification of beneficiaries has come into conflict with the third policy, established by Congress in the phrase following: "...the administrator shall at all times, in disposing of electric energy generated at said project, give preference and priority to public bodies and cooperatives." Except in Washington, IOUs serve a majority of the domestic and rural customers, and the conflict between their interests and the preference for public power has become pivotal.

The fourth policy, a restriction on wholesale power contracts to twenty years' duration has hobbled thermal power plant financing. Bonneville's mandate as a marketing agency meant that BPA could not *purchase* power, but could only sell power produced in federal projects. This means that BPA cannot build or operate--or contract to have built--thermal plants. Marketing authority has thus assumed an importance out of proportion to its economic significance. Thermal plants have economic lifetimes of roughly thirty to forty years. These service lives are paralleled by longer amortization periods for the recovery of the initial capital investment. If BPA is to serve as the linchpin of regional power financing--as many have argued it should--power sales committing BPA beyond the twenty-year limit set by Congress will be required. If there were no controversy, this limitation would probably be readily changed.

The policy of public preference lies today at the center of debate over the future of regional power development. Until the early 1970s, there was a surplus of hydropower, even after public agencies met their market demands; the preference policy then affected planning for the future, not allocation of power in the present. Now that Columbia hydro system is virtually complete, the preference policy means that publicly owned utilities will divide the low-cost power among themselves. The price differentials to retail customers can be significant, as shown in Table 4.

Preferential treatment of publicly owned utilities is not, however, easily changed. Public preference appears in a variety of federal statutes, congressional documents, and legal interpretations dating back to 1906. In the Reclamation Act passed that year, Congress declared that surplus power from federal irrigation projects should be provided so as to favor "municipal purposes." A number of succeeding pieces of legislation reaffirmed this commitment, notably the TVA Act of 1933, which mandated preferential treatment of "states, counties, municipalities and cooperative organizations of citizens or farmers" in the sale of power, installation of transmission lines, and extension of credit for power development.[50] The Bonneville Project Act itself has four statements of public preference.

After the Depression, however, the tide began to turn, as the growth of the investor-owned sector and a general reluctance to extend government enterprise strengthened the voices of public power opponents. By 1954 preference language proposed by Oregon's Senator Wayne Morse was voted out of the Falcon Dam Act. Since then Congress has not written public preference into any power legislation. Even so, the existing preference policy declarations have been repeatedly upheld in judicial and administrative interpretations. Indeed, in 1967 the solicitor of the Department of the Interior held that Indian tribes functioning as local governments qualified as preference customers for the purposes of allocating federal power, thus extending the scope of the preference policy. It is clear that public preference is a policy backed by precedent and political interest. Changing it would entail public debate ranging far beyond the Pacific Northwest and engaging far more than merely technical concerns.[51] The divisions it fosters nonetheless complicate the region's power situation in important, sometimes pivotal ways.

In the 1930s and '40s the preference policy was the mainstay of the public power movement. With Bonneville providing encouragement and analytical support, investor-owned utility assets were taken over by enforced public purchase and new rural markets were opened up by cooperatives and public utility districts. Under its power marketing authority, BPA built transmission lines to serve new public customers with low-cost Columbia River hydro. Public power made much larger inroads in Washington than elsewhere in the region; other states' laws made the organization of publicly owned utilities more difficult and public support for New Deal progressivism was not as strong. But wherever a public agency claimed power, BPA reserved Columbia hydro for it, seeking through its support of public power to provide the economic stimulus that, New Dealers argued, the investor-owned utilities were too cautious to provide.[52]

After World War II, public power encountered national political obstacles: first, Congress turned sharply conservative, and then the election of Dwight Eisenhower as President cooled the ardor of the federal government for social planning and intervention. The political constraints proved, however, to be less influential than technology and economics. As we shall discuss in the next section, the growing scale of the regional power system made coordinated operation of generating projects imperative, whatever their ownership. By the 1960s, the ideological distinction between public and private power seemed to be on the wane;[53] the problem of allocating low-cost federal power has reignited dormant arguments.

Two other legacies of the public power era, one technological, the other economic, remain influential today. First, the harnessing of the Columbia River required not only dams but also a way to transport the electricity they produced. In the long run, the most widespread effects of federal investment are likely those due to the construction, beginning with the creation of BPA, of a region-wide high-voltage transmission grid. The regional grid has provided enormously important economies of scale in power planning and marketing. The original purpose of the grid was to spread the benefits of federal dam development throughout the Northwest, especially to rural areas.

Second, the very use of dam construction and other public works for *regional* development was a key premise of the New Deal's attempts at restarting the Depression-stalled economy. Thus Bonneville carried out a "postage stamp" pricing policy: the cost of bulk power was uniform throughout the grid area. This policy was meant to encourage industrial location throughout the region, particularly to promote development of rural areas. Today, it stands as a reminder that cost-of-service pricing, the utility economics norm, is not sacred when "higher" purposes supervene.

Coordination and partnership

BPA was born in the public power movement, but many of its formative experiences came during World War II. In 1942 the newly organized War Production Board of the federal government was faced with increasing demands for war-related industrial output and electric power in the Pacific Northwest. The director of the board's utilities branch urged BPA to provide a link to Washington Water Power, an investor-owned utility, in order to increase production in its service area. Thus was born the Northwest Power Pool, a joint operating and planning agreement among the region's utilities, both public and private, built around the exchange and sale of bulk power transmitted via the federal grid.

The war also sped the growth of the direct-service industrial customers, as requirements for aircraft led to soaring demand for aluminum. Despite the urgency of wartime production, BPA sought to infuse social purpose into the rush of industrial development. Alcoa's plan to build a gigantic pig-aluminum plant near Spokane was successfully opposed by Bonneville, and the increased aluminum needs were met by three smaller plants at three different locations. The economic benefits were spread more widely, and Alcoa's monopoly power in the aluminum industry was weakened when two of the plants were built by competing firms.[54] BPA's success at meshing private enterprise and public purposes also underscored the important role of the DSIs in regional power affairs; it is a role which has remained large.

Toward the end of the war demand for aluminum slackened, and BPA, along with other business and political interests, lobbied anxiously for government stockpiling of aluminum ingot, to prop up lagging sales. But the American economy surprised the prognosticators. When the war ended vigorous economic growth continued, and the longest boom in American economic history was launched; the Depression was over, not to return. The demand for aluminum picked up smartly.

The formation of the Northwest Power Pool under the pressure of the war marked a moderation of the aggressive public power policies of the New Deal, as IOUs were recognized as partners with Bonneville and the publicly owned utilities, for technical coordination at least. Wartime sales of federal power to IOUs grew so large that they could not be abruptly halted in 1945. And the enthusiasm for taking over investor-owned assets abated as well. With President Roosevelt's death in 1945 and the election of a Republican-dominated, conservative Congress in 1946, the federal commitment to public power waned. Funds for BPA's economic and regional planning program were cut by the House Appropriations Committee and, without the financial independence of TVA, Bonneville had no choice but to curtail its activism on behalf of public power.[55]

The mood of the nation remained favorable to business, an inclination reflected in the election of Dwight Eisenhower in 1952. In his inaugural address President Eisenhower announced the abandonment of public power as a federal policy:

> The best natural resources program for America will not result from exclusive dependence on Federal bureaucracy. It will involve a partnership of the states and local communities, private citizens and the Federal Government, all working together. This combined effort will advance the development of the great river valleys of our nation and the power that they can generate.[56]

The substitution of partnership for the confrontation implicit in the promotion of public power marked a fundamental shift in the institutional relations governing the development of the regional power supply system. For in the slender reference to "private citizens" Eisenhower legitimated full participation by the investor-owned utilities in power planning. The cooperation begun at the technical level in the Northwest Power Pool now extended to investment in new facilities. But what the President could grant through federal *administrative* policy was constrained by federal *legislative* policy: the preference clause remained in the Bonneville Project Act, and in other legislation as well, entrenched and protected by Congressional powers.

Under Eisenhower, partnership meant withdrawal. The national government would no longer support the publicly oriented regional development process. The executive branch did not seek, and Congress did not appropriate, funds to build more dams in the Northwest, though projects then underway were completed. Demand continued to grow, however. The IOUs were anxious to carry on with expansion of Columbia River hydro, since the public preference established by law meant that the IOUs would lose access to low-cost federal hydropower once public agency demands grew to consume the available supply.

Thus, beginning with the construction in 1957 of Priest Rapids Dam, five hydropower projects were built on the main stem of the Columbia as joint ventures between IOUs and public power agencies. The partnership policy took explicit form. Public agencies owned and operated the dams, financing them at the low interest rates available for local government tax-exempt bonds. But the dams were paid for through long-term contracts that sold power to IOUs. It was a precedent-setting arrangement: public financing provided economic advantages while the private sector provided capital.[57]

The building of these five mainstem dams on the Columbia essentially harnessed the hydro potential remaining in the American reaches of the river. Yet the partnership policy had institutional quirks: a substantial quantity of hydroelectric energy remained to be captured through further developing the upper Columbia basin, lying mostly in Canada.[58]

In brief, the demand for power is much more steady the year round than is the supply of water for hydro generation. Although "run of the river" hydro projects like Bonneville Dam do impound some water behind them, the storage capacity of these projects is too small to regulate river flow over the year. Power is therefore lost during the spring and early summer, when river volume is high from melting snowpack and water must be spilled over the dam. Further, hydro capacity during the low-flow season is limited by the amount of water in the reservoirs. Storage dams, which do not themselves contain generators, can thus enhance the production of hydropower at downstream projects through releases during the winter when natural flow is low.

The partnership policy had encouraged non-federal development of the Columbia, but the legal framework of the Federal Power Act made non-federal storage projects especially difficult to launch. As a result, there was an emphasis during the partnership years on building run-of-the-river hydro projects, leaving the Columbia system with generation facilities that could not be utilized fully for meeting base load demands.

Four interlinked agreements were achieved in 1964, paving the way to completion of the hydroelectric system.[59] The keystone was the Columbia River Treaty, ratified in 1964 by the Canadian government. The treaty provided for cooperative development of the upper Columbia, and apportioned the economic benefits that resulted. Canada was to build three storage dams, Mica, Arrow, and Duncan; Canada also agreed to United States construction in northwestern Montana of Libby Dam, whose reservoir floods land forty-two miles into Canada. These projects have increased upstream storage capacity by 15.5 million acre-feet, lengthening the "critical period"[60] from 7 to 43 months. By providing winter releases of stored waters, the treaty dams enhance the productive capacity of hydro projects downstream by an estimated 1,900 average megawatts.[61]

This capacity could only be fully utilized, however, if the American hydro facilities were operated in harmony. Accordingly, the Pacific Northwest Coordination Agreement of 1964 was signed by the region's utilities.

This constituted a second major institutional shift. The agreement had been negotiated earlier, as growing power demands in the Northwest had made apparent the need to maximize hydro production with existing facilities. The treaty with Canada served to formalize an interim agreement already in place on an experimental basis. Forbiddingly technical in its terms, the Coordination Agreement effected a consolidation of the operational control of hydro facilities, decreasing the day-to-day autonomy of the participating utilities but increasing their productive capabilities.

The third institutional arrangement of 1964, the formation of the Columbia Storage Power Exchange (CSPE), was instrumental in financing Canada's participation in the development of the upper basin. Half the power benefits provided by the Canadian storage dams were assigned to Canada. Yet in the 1960s there was insufficient demand in western Canada for this power. It was sold instead through a thirty-year contract to a consortium of Pacific Northwest utilities, organized as CSPE, a nonprofit corporation marketing treaty-derived power. But the Northwest utilities could not use the power immediately either, though they did anticipate that Northwest demands would materialize within thirty years. CSPE accordingly sold the treaty-derived power on shorter term contracts to utilities in the Southwest, mainly in California. CSPE financed its purchases with a private bond offering, paying the Canadian government in advance for thirty years' worth of power benefits. Canada used these funds to build the three dams required to meet its treaty obligations.

The final piece of this complex undertaking was the building of a high-voltage transmission network needed to transport CSPE power to the Southwest. The Pacific Northwest-Pacific Southwest Intertie was authorized in 1964 by Public Law 88-552 as a joint venture involving federal, IOU, and public power participation. (See Figs. 9 and 10.)

Like the Coordination Agreement, the intertie had been under discussion for some time, and for purposes beyond conveying treaty power. But the preference policy had created two issues which had blocked a final agreement. Once the intertie was built, economic transmission to the large power markets of the Southwest

Figure 9. *Pacific Northwest-Pacific Southwest Intertie.*

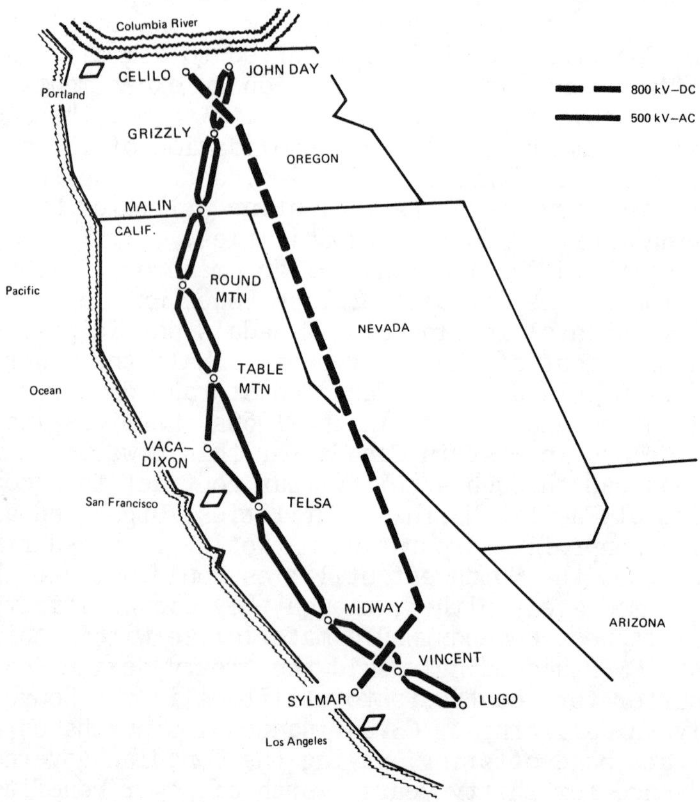

Source: *Pacific Intertie* (Portland: BPA, 1977).

would be available. The policy of public preference seemed, then, to give Southwest public agencies a claim on the Northwest's hydropower. This caused concern especially to Northwest IOUs and direct-service industrial customers. But public power advocates were worried as well. In part because of the partnership policy, planning and financing of the intertie had involved the IOUs extensively. California's Pacific Gas and Electric Company, in particular, would own a major segment of the line. On the defensive because of the partnership policy, public power interests feared that

Figure 10. *Interregional Interconnections of the Western Systems Coordinating Council.*

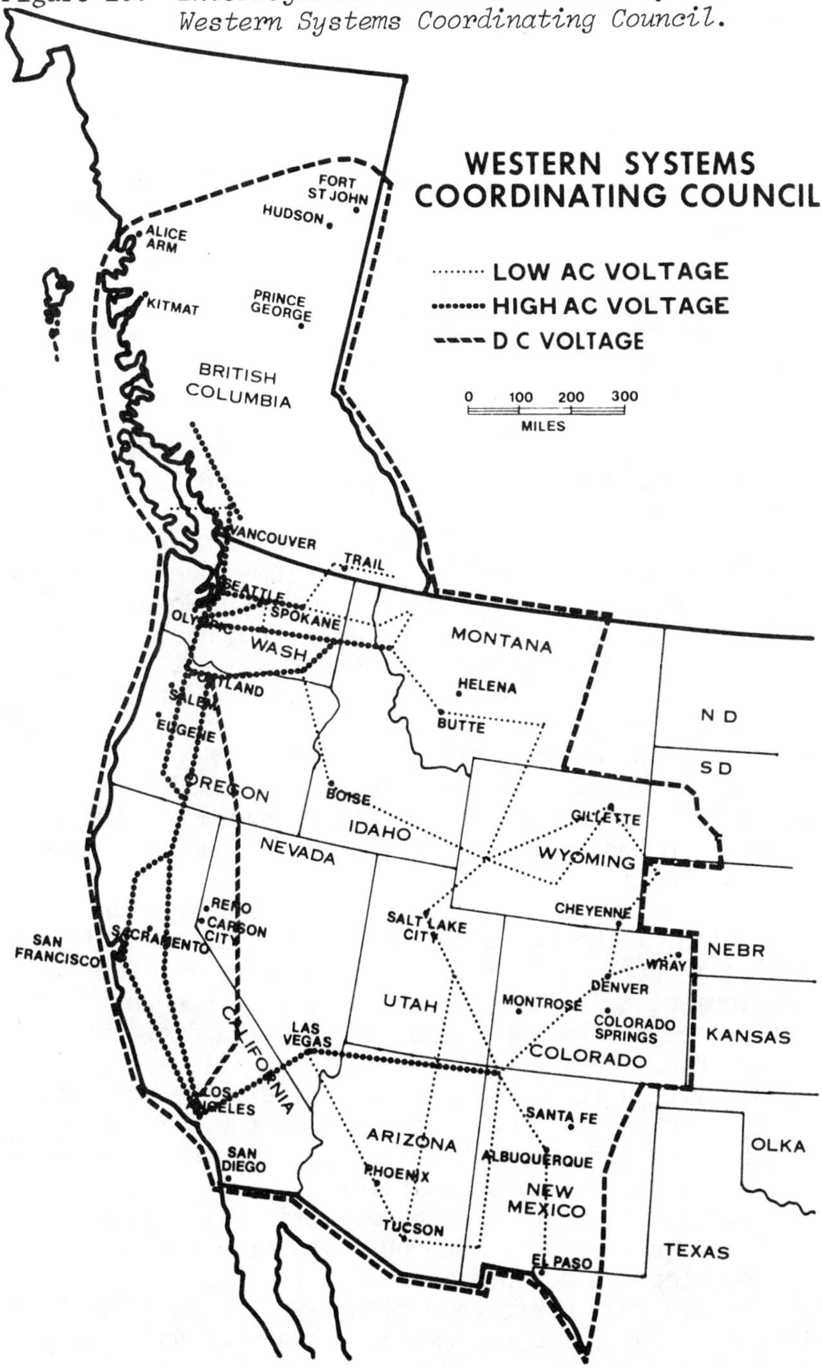

Source: Western Systems Coordinating Council brochure (n.p., n.d.)

PG&E's control of part of the intertie would upset the balance between public and private power in the Southwest.

Both concerns were allayed in the language of P.L. 88-552. A regional preference clause was inserted in the bill, giving Northwest utilities rights to power ahead of the claims of others outside the region; this would keep Bonneville's hydro "home." And the intertie, although financed in part by investor-owned utilities, was declared to be a common carrier, so that access to transmission would be protected.

The development of the upper Columbia basin and the construction of the intertie completed the regional hydro system. The regional development program initiated by the construction of Bonneville Dam in the 1930s had been achieved.

The original federal commitment to public power in the 1930s was moderated by the wartime necessities which led to the organization of the Northwest Power Pool. The ending of the Depression-induced willingness to use government for economic development brought a long period in which power planning was fragmented and the IOUs gained parity with the public power agencies. The IOU position was further strengthened by the controversy over the New Production Reactor, to be discussed in the next chapter. And the completion of the hydro system coincided with the start of interregional power exchanges via the Pacific Intertie.

It is interesting that the institutional and policy evolution of the hydro system has been shaped mainly by shifts in national politics and policies for regional development. The mission and design of BPA were strongly influenced by the hopes and fears surrounding the TVA.[62] As an advocate of public power, Bonneville was a federal agency with a regional mission. Later, the national mood of conservatism dictated the partnership policy, rather than any regional reluctance to proceed with the harnessing of the Columbia River.

With the events of 1964 a distinctly regional activism became apparent. The benefits of the upper Columbia development accrued to the Northwest, most visibly in the regional preference clause in the intertie legislation. Bonneville had emerged as a regional agency with a federal pedigree--and federal constraints. The succeeding phase of electric power development would reveal how confining the institutional heritage of hydro could be.

CHAPTER III
Hydro-Thermal—The Second Generation

> The object of the system, [Henry Ford] said, was to fill "needs the public [is] not yet conscious of."[1]
>
> --Elting E. Morison

Since the late 1950s electric power planning in the Pacific Northwest has been dominated by a technological idea of enormous influence, the hydro-thermal concept. The harnessing of the economically realizable Columbia River hydropower was expected by the 1970s. The Northwest would then have to turn to other, more costly supplies of electricity. All of these sources transform heat into electricity and are known collectively as "thermal generation." For technical reasons, hydropower is an excellent complement to thermal generation; the system containing both hydro and thermal components promised to be far more supple and substantially more economical than thermal alone. The hydro-thermal dream was not without costs and disadvantages, but its technical appeal induced far-reaching institutional changes among Pacific Northwest utilities.

These institutional shifts are discussed here as they emerged from two major decisions. First, we examine the political wrangle that led to the first nuclear power plant contributing electricity to the BPA grid, the New Production Reactor at Hanford, Washington, authorized by Congress in 1962. The debate over the N-Reactor, as it is known, crystallized a commitment to a regionally integrated power supply. This commitment was underscored in 1964, with the enhanced coordination and regional preference policy precipitated by the Columbia River Treaty discussed in Chapter II.

The urgency of thermal power seemed great during the rapid growth of the 1960s. The second decision we shall examine is the formulation in 1968 of the Hydro-Thermal Power Program (HTPP). Modified in 1973 by its utility sponsors, HTPP is what now remains of an official regional power plan. It is "in shambles," as one of its proponents mourned in 1976. The struggle to rehabilitate--perhaps to redefine--HTPP remains the central issue in regional power planning.

The technological and financial imperatives given concrete form in the Hydro-Thermal Program reflect shifts in economic, organizational, and political powers. We suggest below that Galbraith's analysis of the "new industrial state" aptly describes the institutional mantle of the hydro-thermal generating concept.

THE N-REACTOR AND PUBLIC-PRIVATE COOPERATION

By the time Senator Henry Jackson proposed nuclear electric power for the Northwest, the region had long housed a major component of the atomic energy establishment.[2] The plutonium used in the first American nuclear weapons was produced in a reactor built in 1943 near the once-obscure town of Hanford, Washington, close to the confluence of the Columbia River and its principal tributary, the Snake. The Hanford site was chosen in part because its isolation and obscurity facilitated military secrecy. It also had access to rail transport, Columbia River water needed to cool a nuclear reactor, and abundant electricity from the Bonneville grid.

A decade after the Second World War gave way to the Cold War, the Hanford Atomic Works, administered by the U.S. Atomic Energy Commission (AEC), had grown to include eight reactors producing fissile materials for military purposes. The surrounding towns of Richland, Pasco, and Kennewick had prospered, and proclaimed themselves the Tri-Cities.[3] In 1956 Senator Jackson, a member of the powerful Joint Committee on Atomic Energy, recommended the construction of a ninth plutonium manufacturing facility, the New Production Reactor. The

N-Reactor, Jackson argued, presented the opportunity to advance civilian as well as military ends, for it was designed to generate electricity as well as plutonium. It took six years to harmonize these dual purposes.

Plutonium is a man-made element, one of whose varieties, plutonium-239, is used in both weaponry and power generation. Plutonium is made from uranium, a lighter element, by bombarding uranium with neutrons in a reactor. The neutrons, in turn, are produced in a controlled chain reaction which splits, or fissions, uranium nuclei. These fission reactions produce heat as a by-product. The process of making plutonium closely resembles the fission process by which commercial nuclear plants generate steam to power generators. Indeed, the heat produced by Hanford's plutonium plants required quantities of Columbia River water to dissipate.

The technical opportunity to combine plutonium and steam production had political possibilities as well. The Eisenhower partnership policy prevailing in 1956 left little room for federal participation in the development of the Columbia River basin. But military superiority, especially in nuclear affairs, was a federal responsibility that transcended partisan rivalry over public power--or so Jackson urged. The N-Reactor was a way of continuing the federal support of Northwest economic development that the partnership policy eschewed. In accord with the precedents established in river basin development, the N-Reactor's electricity was to be provided preferentially to public power agencies.

Jackson's support of public power was well known, however, and in July 1956 the House of Representatives, though controlled by Democrats, rejected the N-Reactor proposal. The Eisenhower Administration and its Congressional allies feared that Jackson's "bill might be a step toward converting the AEC into a public power-for-sale agency," as Republican conservative Senator Bourke Hickenlooper put it.[4] These suspicions combined with the opposition of coal state Congressmen, led by Representative James Van Zandt, a Pennsylvania Republican, who feared that cheap nuclear energy would further depress that declining industry.

Two years later the N-Reactor was funded by Congress and reluctantly approved by President Eisenhower. The principal spur was the Soviet *Sputnik*, launched in

October 1957, which cast a pall over America's technological self-confidence--and mobilized enormous political support for U.S. science. Moreover, the power generation aspects of the project had been sharply curtailed. Although the reactor's design would permit dual-purpose operation, the generating turbines and associated equipment were eliminated from the project. Even so, Eisenhower criticized the N-Reactor as inappropriate federal meddling in private electrical generation. But the AEC appropriations bill included many other, less dispensable monies as well, and Eisenhower signed it.

When the Kennedy Administration took office at the beginning of 1961, Northwest nuclear advocates tried anew to secure funds for dual-purpose operations at the N-Reactor. But the coalition of private power and coal-state interests once again prevailed, and a $60 million appropriation was deleted by the House of Representatives in September 1961. The IOUs' opposition was led by the Edison Electric Institute, a national lobbying organization. One analyst of the N-Reactor controversy, historian Bonnie Pendergrass, concluded that the national lobby may not have represented Northwest IOUs:

> ...regional loyalties and self-interest were undermining traditional ideological opposition to public power as "Giant Power" technology increased construction costs and made interconnection...more practical.[5]

As Pendergrass points out, the region's IOUs were not vocal opponents of the N-Reactor. Yet the pattern of public-private cooperation remained to be worked out. In particular, the joint financing of hydropower projects, hammered out in the 1950s under the partnership policy, had not been recognized as the key precedent for thermal power plant development.

Enter WPPSS

Instead, when Congress rejected the N-Reactor generating system, its proponents turned to state government--and brought onto the hydro-thermal stage an organization whose importance in regional power planning was destined to mushroom. Nine days after the House vote thwarting the N-Reactor power project, Washington

Governor Albert Rosellini ordered a study "to investigate the feasibility of the State undertaking the installation of electric generating facilities at the Hanford convertible reactor."[6] The interagency inquiry which followed found a feasible state agent in the Washington Public Power Supply System (WPPSS, pronounced "wupps"), headquartered in Richland, at the boundary of the Hanford reservation.

The Supply System was born in 1957, a child of the public power movement, conceived in response to the partnership policy.[7] It is a "joint operating agency" chartered under Washington State law, a consortium of public power agencies able to buy, build, operate, and own generation and transmission facilities. Founded by seventeen public utility districts, WPPSS now has a membership of twenty-two agencies.[8] The joint operating agency, roughly analogous to a utility holding company in the corporate world, plays an instrumental role in public utility finance. WPPSS has the legal power to raise funds through the sale of securities, revenue bonds in particular. Income derived from securities issued by state and local governments is not subject to federal income tax. This traditional inducement to investors to support local government finance lowers the interest rate paid by these government borrowers.

WPPSS thus shares with its member utilities a powerful financial advantage over the IOUs, which have to raise funds at higher corporate market rates; the Supply System's revenue bonds even enjoy tax advantages compared to revenue bonds issued by federal agencies such as the Tennessee Valley Authority, whose yield is subject to federal income tax.[9] In addition, by aggregating the expected needs of its members, WPPSS can invest in larger facilities, taking advantage of economies of scale. Though it was formed too late to participate in the five non-federal hydro dams built during the Eisenhower years on the main stem of the Columbia, WPPSS was ideally positioned to finance the generating component of the N-Reactor in 1962.

As arrangements for WPPSS to sponsor the N-Reactor project neared completion in the summer of 1962, the General Accounting Office--the audit branch of Congress--issued an advisory legal opinion. GAO ruled that contracts between the Supply System and the Atomic Energy

Commission, owner of the N-Reactor itself, required congressional approval. The contest between public and private power was joined once more.

Jackson's Compromise

On July 11, 1962, during a hearing of the Joint Committee on Atomic Energy, Senator Jackson suddenly moved to break the deadlock. He suggested that power produced by the N-Reactor should be shared equally between public agencies and Northwest IOUs. Although coal interests fought back, their alliance with privately owned power companies had been undermined. A determined floor battle waged by Northwest Congressmen late in the summer of 1962 carried the day. By autumn President Kennedy had signed the bill authorizing a dual-purpose N-Reactor, and nuclear-generated power began flowing into the Northwest Power Pool grid in April 1966.

Jackson's dramatic compromise came, as the senator recalled it, "right off the top of my head...I didn't brief anybody because I hadn't given it that much preparatory thought."[10] Public power advocates present at the hearing, including Supply System president Owen Hurd, were taken by surprise. They worried that the sharing of N-Reactor power would dilute the public preference policy, already weakened by eight years of Republican "partnership." But they did not object: they trusted Jackson's political judgment and feared that the nuclear project would be lost once more without a major concession to the IOUs.

Despite the improvisation in his statecraft, Henry Jackson had broken more than an ideological impasse with his proposal to share the benefits of federal investment between public and private owners of electric utilities. For the key to financing the large-scale development of thermal power in the Pacific Northwest was--and remains --a three-cornered collaboration among private and public utilities and the federal government. The N-Reactor compromise preserved the spirit of Eisenhower's call for partnership between public and private investors, but it gave the Republican penchant for a lower federal profile a reverse twist. Henceforth, the federal government would lead and coordinate the partnership, rather than simply withdrawing in order to force cooperation upon reluctant public power agencies.

Thus a regional commitment to thermal power development began to take shape. WPPSS, an organization representing the common interests of public power agencies, worked at a financial scale and level of technical sophistication comparable to those of the IOUs. More important, the ideological gulf that separated public power from the private electric companies seemed to be bridged by the successful sharing of the N-Reactor's output. The public-private rift seemed to heal further as the Columbia River hydro system reached maturity with the upper basin developments launched in 1964. The next step was full commitment to thermal power: the federal government, through the Bonneville Power Administration, was ready to lead the way.

THE HYDRO-THERMAL PROGRAM

Introduction of thermal plants into the Northwest was officially discussed for the first time in the 1958 Army Corps of Engineers "308" Review Report.[11] The Corps noted that if growth in electrical power demand materialized as projected, the economically feasible hydro projects in the Northwest would be constructed by about 1975. This would bring to an end the first stage of regional electric power development, the hydro phase. The second, hydro-thermal stage of regional growth, running to the end of the century, would be characterized by the addition of thermal facilities to the regional supply system (Fig. 11). The 1958 report also talked hazily about a third stage of development, one in which thermal generation would come to dominate the regional supply system.

The hydro-thermal phase of evolution could take advantage of a technical difference between thermal and hydro generation. Thermal plants are basically large pots of boiling water. Adjusting their output takes a good deal of time, since steam pressure takes some minutes to change even when a plant is running. In nuclear plants changes in power output, which necessitate changes in operating temperature, cause delicate fuel rods to bend and bow, threatening their ability to contain their highly radioactive contents. By fortunate coincidence, a hydro dam has complementary technical characteristics.

Figure 11. *Projected Additions to Electric Energy Supply in the West Group Area.*

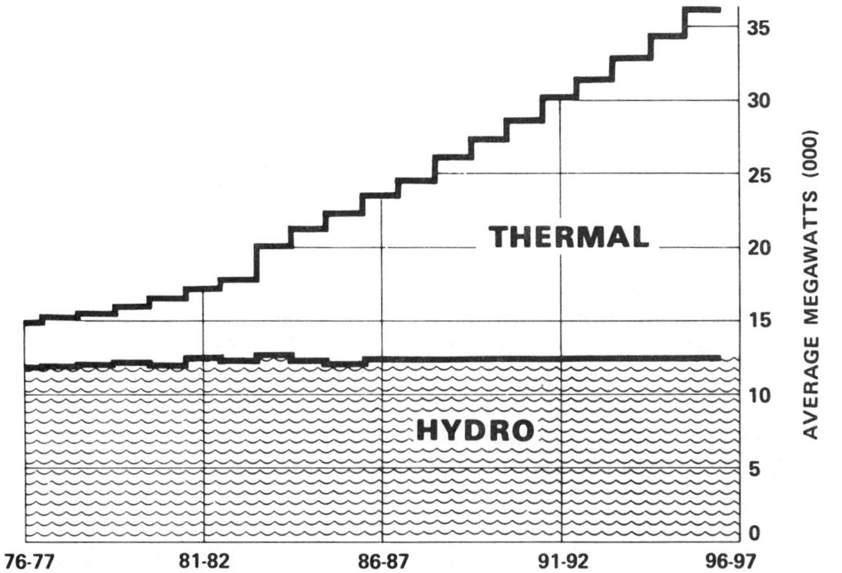

Source: *The Electric Energy Picture in the Pacific Northwest* (Portland: BPA, 1976).

The flow of water can be adjusted within seconds, as penstocks are opened or closed. (The environmental implications and multiple-use conflicts inherent in this use of the hydro system were not recognized until much later.)

The hydro-thermal concept promised to be an innovation in utility economics as well. Power generating systems are traditionally built to reflect the fact that electricity cannot be stored, but must be generated as it is demanded. Demand varies substantially over the course of the day, usually rising sharply in the morning and subsiding during the day, then climbing again to its daily maximum around suppertime, followed by the decline to its minimum late at night. Power demand also changes with the seasons, as power use for heating, cooling, and --of increasing importance--irrigation varies. In conventional utility operations, the cheapest generating technology is used for the base load: the constant, around-the-clock supply of power needed to supply the late-night demand. As demand rises above baseload, successively more expensive generating methods are called into play, until peak load is met with the costliest supply in the system.[12] Thus, to meet early evening peak loads a major utility in the Northeast will use gas turbine generators, which have very high operating costs, but are used for only an hour or two each day.

The hydro-thermal system projected for the Northwest is fundamentally different. Thermal plants are run as baseload supplies, while hydropower output is adjusted to meet varying demands during the course of the day. This is done even though thermal power is costlier than electricity from hydro dams. The reason, as indicated above, is partly technical. But it is largely economic: the flexibility of the hydro system makes the hydro-thermal mode of operation the least costly way of expanding the power supply. Although thermal plants are more expensive to run than the dams, far the greatest part of their expense--between 50 and 85 percent of the cost of production--is due to fixed capital, money tied up in financing plant construction in the first place.[13] Capital costs accumulate whether the thermal plant is operated or not, since their repayment is essentially the paying off of a loan. Thermal plants are accordingly most economical when run at constant output. By comparison, both capital and operating costs of the hydropower dams are low: the dams themselves have been partly paid for in the years the hydro system was being developed, and there are no fuel charges. The hydro-thermal concept allows the flexible hydropower network to be operated so that the thermal plants are built as late as possible, to delay their capital costs, and, once built, are paid for as rapidly as possible, to minimize interest charges.[14]

Institutional Design

Late in 1966 BPA administrator Charles Luce appointed a special task force to study thermal power additions to the regional grid. The next year Bonneville convened the Joint Power Planning Council (JPPC), an organization with representation from BPA and 180 of the utilities in and bordering the West Group area. Chaired by Luce, JPPC was a body designed to bring together Pacific Northwest utilities. It issued plans for the Hydro-Thermal Program in 1968. HTPP was only a first step toward a regionally integrated power system, but it was a fateful one. The hydro network had been dominated by Bonneville's role as leader of the federal system. It would not be clear who could or should lead the much tighter coordination demanded by thermal technology.

As the Northwest's demand for electricity grew beyond the rivers' capacity, thermal generation would come gradually to overshadow the mighty dams that had brought civilization to the inland desert, and industry and jobs to Portland, Boise, and Seattle. Thermal power on a large scale was developed first by investor-owned utilities, whose access to federal hydropower was curtailed by the growth of public power demand under the preference policy. Thus, in 1971 and 1972, a consortium led by Pacific Power and Light and Washington Water Power built a two-unit coal-fired plant near Centralia, Washington, adding 1,400 megawatts to the regional supply.

But reliance on private enterprise would not be enough for the juggernaut demand that seemed to impend. The literature describing HTPP contained projections drawn out to 1990, though the program itself was scheduled to last barely till 1980. By February of 1990, the projections suggested, there would be 21,400 megawatts of thermal power added to the Northwest supply--twenty nuclear and two coal-fired plants--more than doubling the 10,600 megawatts of firm hydro capability available in 1970.[15] Thermal power seemed inevitable: growth in the 1960s seemed unstoppable.

Since then, the headlong momentum of exponential growth has slowed, but it has not stopped by any means. The medium-growth scenario of the Northwest Energy Policy Project (NEPP), published in early 1978, is substantially more modest than the utilities' forecast. The NEPP study extrapolates electricity demand at somewhat less than 3 percent per year; still, by the year 2000, this implies thermal power resources of over 25,000 megawatts in the Northwest, more than double the hydropower supply.[16]

Such magnitudes were awesome, even intimidating, to utility men unfamiliar with thermal technology. "I had people say to me, 'I don't know how you can build that many plants,'" recalls Don Hodel, who was appointed Bonneville deputy administrator in 1969, just as HTPP was launched.[17] Those least familiar with thermal power were the dozens of small public agencies serving rural communities, agencies born from the largesse of the Columbia River and the feisty determination of public power enthusiasts. According to Bernard Goldhammer, then power manager at BPA, the rural cooperatives "would have preferred not to be involved, really in the question of power supply....But they would continue to buy from Bonneville."[18]

Indeed, the tumultuous heritage of the public power movement meant that any regional power development program had to be a political as well as a technical creation. After the ups and downs of the hydroelectric era, Bonneville's relations with its public utility clientele had all the internecine characteristics of a family with adolescent children. The smaller publics, realizing their continued dependence on BPA and the federal system, sought to make their presence felt through the Public Power Council, formed in 1969. The larger public agencies, such as the Eugene Water and Electric Board, sought to maintain their independence. The issue, noted Goldhammer, was managerial autonomy: "The control is obviously in the hands of who has the generation....If municipal systems had their own generation, they would be completely free."[19] But the logic of regional development meant that control would need to be shared.

For the industrial customers and investor-owned utilities, the Hydro-Thermal Program posed large and uncertain costs. As the demand for federal hydropower mounted, the surpluses which they had purchased for decades through firm contracts at bargain rates would disappear. HTPP offered a way to preserve some of the advantages of hydro for industrial and IOU customers, but only if the concept of a coordinated regional program could be brought to fruition.

Given the legacy of bitterness and suspicion, together with the substantial uncertainties of thermal power, HTPP was drawn modestly: "You could never get agreement on any long-term program," Goldhammer felt.[20] So HTPP took shape as a ten-year program involving the construction of seven thermal plants (Tables 7 and 8).

Table 7. *Capital Investments Projected in the Hydro-Thermal Power Program, 1971-81.*

INVESTMENT	PROJECTED COST ($-millions)
Non-federal share: $4.6 billion	
Thermal Generation	2,072
Hydro Generation	3
High-Voltage Transmission and Distribution	2,300
Federal share: $2.5 billion	
Hydro Generation	1,300
High-Voltage Transmission	1,240
TOTAL	7,135

Source: *The Hydro-Thermal Power Program: A Status Report* (Portland: BPA, 1971), pp. 21-22.

The federal government would augment the hydropower capabilities of the Columbia River system, to provide additional peaking capacity, and it would also expand the regional high-voltage transmission grid. The utilities would build plants and extend their distribution systems for delivery of power to a growing retail market.

The long-term implications of continued growth were insistent, and BPA reminded the utilities of their scope by projecting hydro-thermal development out till 1990 (Fig. 12). But the central financial structure of the Hydro-Thermal Program--the procedure of net billing--"was developed because it had a self-contained limitation," Hodel remembers.[21] The American tradition of limited government was manifest. As we shall argue below, the technical and economic logic of a regional scale program points implacably toward centralization. The decentralized, cautious design of HTPP tried to compromise with this logic, but the centrifugal forces inherent in the compromise could not be managed: one

Table 8. *Thermal Power Plants of the Hydro-Thermal Power Program.*

Plant No.	Project	Fuel	Capacity Unit No.	Capacity MW	Date of Operation[d]	Sponsors	Percent Net Billed	Estimated Investment Cost, 1971 ($-millions)
1	Centralia	Coal	1	700	Sept. 1971	PP&L Co. &		242
			2	700	Sept. 1972	WWP Co.		
2	Trojan	Nuclear		1,130	Sept. 1974	PGE Co.	30	232
3	Jim Bridger	Coal	2	500	Sept. 1975	PP&L Co	0	228
			3	500	Sept. 1976			
4	Hanford No. 2	Nuclear		1,100	Sept. 1977	WPPSS	100	352
5	Not Identified[a]	Nuclear		1,100	Sept. 1978	WPPSS		268
6	Not Identified[b]	Nuclear		1,100	Sept. 1979	WPPSS		361
7	Not Identified[c]	Nuclear		1,100	Sept. 1980	WPPSS		389
							TOTAL	2,072

a. WPPSS Nuclear Project No. 1, 100 percent net billed, 1250 MW.
b. WPPSS Nuclear Project No. 3, 70 percent net billed, 1240 MW.
c. Pebble Springs 1, 15 percent net billed, 1260 MW.
d. Except for Centralia, these have all been delayed; plants 4-7 have not been completed as of August 1979.

Source: *The Hydro-Thermal Power Program: A Status Report* (Portland: BPA, 1971), p. 43.

sign is that four of the seven plants in Table 8 have yet to be completed, and all four are seriously behind schedule. HTPP was a first step; it could not set the pattern for future steps.

Preference and Reserves

Although promulgated by JPPC, the Hydro-Thermal Program was very much Bonneville's creation. Not only in its fiscal and organizational structure, but also in the process of building the utilities' consent, BPA was a leader, fashioning incentives and organizing negotiations. Indeed, the hydroelectric heritage had left few besides Bonneville capable of realigning the diverse interests in the region. Despite its historical advocacy of public power, Bonneville had supported and implemented the Hanford N-Reactor compromise, sharing federally subsidized nuclear power between IOUs and public agencies. WPPSS, the sponsor of the Hanford generating project, was too much a Washington State and public power institution to assert regional leadership. The IOUs, like WPPSS, had the requisite economic capacity to lead, but lacked the political credibility and interest to assert themselves strongly. The state

Figure 12. *Capital Investment Requirements of the Hydro-Thermal Power Program, 1971-92, as Projected in 1971.*

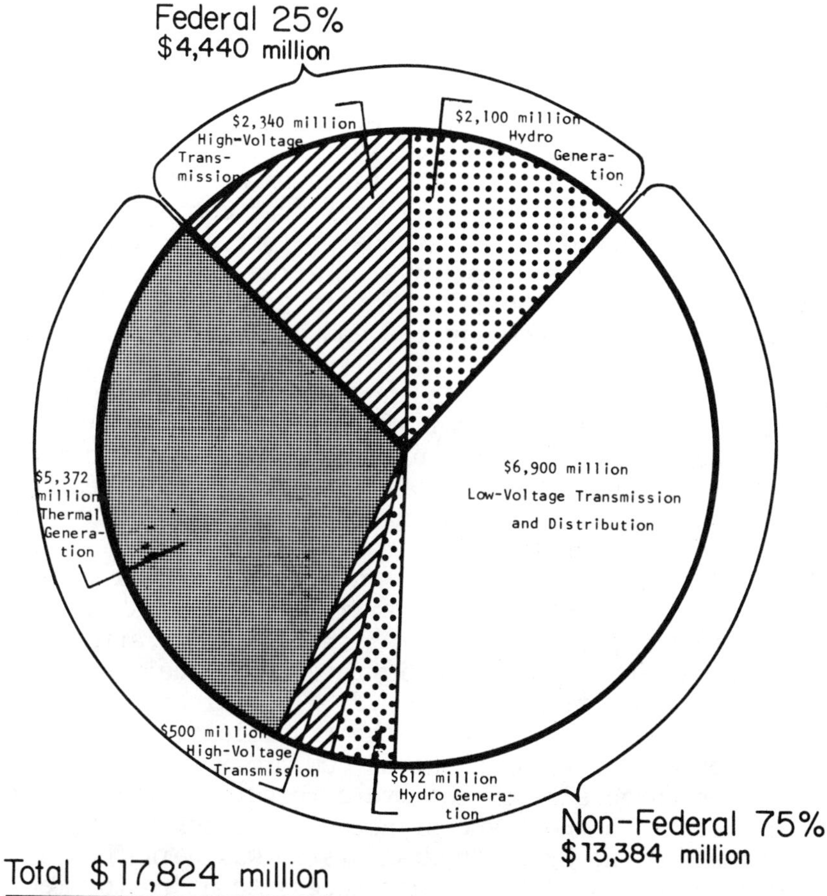

Source: *The Hydro-Thermal Power Program: A Status Report* (Portland: BPA, 1971), p. 23.

governments were inactive: their bureaucracies were content to regulate IOU prices, and regional planning seemed, then, far outside their responsibilities.[22] Perhaps most important, BPA leadership was quietly backed by the Northwest Congressional delegation, especially Senator Jackson, who had by this time risen to the chairmanship of the Senate Interior Committee which oversaw Bonneville.

The Bonneville Power Administration was the central actor of the hydro-thermal drama in another sense, for it was at once the source of the major obstacles to HTPP's design and the means for overcoming those obstacles. The first problem was the preference policy; the second, the financing of thermal plants. The hydrothermal concept called for a fundamental revision of the hydropower system, since all peak load customers would have to be served, not just the preferred public utilities. This required new ways of arranging and using BPA's marketing authority. The institutional design which resulted is complicated and convoluted, a triumph of the corporate lawyer's ability to interpret the law flexibly, and a tribute to the engineer's promise of economic and technological virtuosity.

The policy of public preference entwined the hydrothermal concept like the Gordian knot. But instead of cutting the knot by repealing the preference clauses in the Bonneville Project Act, the planners of HTPP tried to untie it. The hydroelectric heritage vested many interests and historic animosities, and this sharply narrowed the possibilities for institutional arrangements. As former BPA administrator Don Hodel remembers it, "The whole trust of the Hydro-Thermal Power Program ...was to do no violence to the traditions."[23] The preference policy was a sacred cow.

The large size of the hydro resource permitted flexibility on the issue of access. Access to hydropower while the hydro-thermal system was being built could be addressed first, and access to hydro for meeting peak-load demands later, after the system became large enough that public and IOU distributors would have to share it. The first issue could be settled within the bounds of the preference policy. The second could be deferred. Deferral had two benefits; through working together on HTPP, public and IOUs might build enough mutual confidence to permit revision of the policy of public preference; and the economic benefits of hydro would dwindle in importance as thermal power came to dominate costs.

The JPPC planners thus devised a web of obligations and rights which reaffirmed the preference policy while laying the groundwork for moving beyond it. In sales of federal hydro, the public agencies, called "preference customers," were assigned first rights to long-term firm power contracts, providing guaranteed levels of electric power. The IOUs retained their privilege to negotiate short-term firm contracts, and to purchase non-firm power coming available from streamflows above the minimum "critical year" planning target, lower-than-expected demand, and other unplanned sources. Since 1973 no new firm power contracts have been available to the IOUs, though sales of non-firm power have continued (See Table 4).

In support of this two-tiered market BPA supplemented its marketing responsibilities by providing regional reserves. Both public and private utilities were assigned rights to purchase power from the federal hydro resource, if it were needed to compensate for inaccuracies in demand projections, or to compensate for delayed completion of HTPP thermal plants. Reserves were also available to accommodate forced outages of federal hydro generating facilities. These reserves, designed to advance regional ends rather than public preference, helped to regionalize thermal plant investment by spreading the effects of inaccurate forecasts and by providing a federal fall-back in case a thermal plant failed to operate reliably. Regional reserves were in turn marketed to direct-service industrial customers by the IF-1 rate schedule introduced in 1975.[24] In a manner reminiscent of Henry Kissinger's "step by step" Mideast diplomacy, Bonneville sought to unify the once-antagonistic IOUs and publics so that the cooperation and mutual dependence needed for regional power development could grow.

Net Billing

Rights to hydro and reserves did not, however, solve the most important economic problem of regional thermal development: securing financing for the power plants. Although the IOUs had established procedures for borrowing funds to build thermal generating stations, large-scale capital construction was a new and unfamiliar challenge to most of the public agencies. And WPPSS, the builder of three nuclear plants, was a largely untried mechanism for public agency cooperation.

Thermal plants posed difficulties for the public agencies, ironically, because of the publics' historical dependence on federal hydro. The federal Columbia system permitted the development of small public agencies that do not own any generating facilities and rely on BPA for all their electricity. This fragmented market structure inhibited regional cooperation, and, as Bonneville's power manager recalled, "If nothing was done on a regional basis, you would have...the public systems building small plants...at high cost."[25] Not only would costs be high, but the coordination needed to operate the hydro-thermal system would become more complex, and environmental impacts would be needlessly multiplied.

The small size of many public utilities contributed to a related problem: financing large thermal plants was going to be expensive for public agencies. A basic difficulty was lack of equity. As former administrator Hodel points out, in IOU-dominated regions, capital investment is financed through corporate bonds backed by the assets of privately owned electric power companies; in the Northwest, where IOUs account for only half the retail sales of electricity, the private assets available to finance expansion were only half as great, proportionally, as those of other regions.[26] Furthermore, Northwest public power agencies do not own much property because the federal government provides the principal productive assets, the hydro projects and regional transmission network. The public power tradition stressed low electricity prices as a principal goal, and Bonneville had insisted that its preference customers keep retail prices down; as a consequence, the publics had accumulated little capital. The publics accordingly had little to offer lenders by way of collateral. The alternative was to issue revenue bonds--borrowing against the thermal plants' income-earning potential itself. It could be done, but it would be costly; to public agencies used to low-cost federal power, the figures looked discouraging.[27]

Not all the publics were in this position, a fact which complicated BPA's negotiations. Municipal agencies serving major cities like Seattle and Tacoma had invested in major hydro projects on their own, as well as some smaller thermal plants. With their larger, more experienced staffs, together with well-established financing capability in the municipal bond market, they felt capable of managing thermal plants. These public entities resisted Bonneville's leadership, feeling that their autonomy might be compromised by too great a degree of interdependence. They feared, for example, that growth in BPA sales to industrial customers would raise region-wide power costs with no compensating benefits to them or their consumers.

Yet the arguments in favor of regional thermal development were strong, and Bonneville developed a financing arrangement, called net billing, which overcame the reluctance of the public agencies. Under net billing, public utilities participate in the purchase of a thermal plant. On paper, Bonneville averages these thermal costs with its hydro system costs. Though Bonneville never makes a capital outlay for the thermal power, the effect of the net-billing procedure is to put the financial resources and equity of the federal system behind the preference customers in these thermal plant ventures. Net billing was established to provide a bridge for the public systems, from a period of nearly complete dependence on hydropower from BPA to the period when the utilities would have to develop and finance generation entirely on their own. Though net billing has been used in other regional power pools, it was first brought to the Northwest in the Hydro-Thermal Program. Net billing became the linchpin of the program.

The complications of net billing reflect sensitivities growing out of the battles over public power. While Bonneville has authority to market federal power and to transmit power generated at federal and non-federal facilities, it is not authorized to purchase non-federal power for resale, nor to build and operate a power plant without explicit approval from Congress. Purchase of power or ownership of power generation facilities is thought to be too assertive a role to be played by an agency with a public power past.[28] Indeed, opposition to Bonneville purchase authority came not only from IOUs, but also from the larger public agencies, jealous of their independence. As a former BPA administrator summarized the prevailing sentiment,

"Don't let Bonneville buy."[29]

Net billing is a means of purchasing power from non-federal sources--thermal plants--without using money. In place of an exchange of cash, power is purchased with credits developed in the course of Bonneville's marketing of federal hydropower. (See Fig. 13). The accounting begins at the generating station. A public utility agrees to participate in a thermal plant--that is, to pay for a fixed proportion of its output each year. In the example in Figure 13, Sasquatch PUD participates in one-sixth of the Crabshell Nuclear Project (CNP), paying for 100 of its average 600 megawatts output.[30] Thermal plants may be owned by a public entity such as WPPSS, or by a consortium of public and private utilities, as in the case of the Trojan nuclear plant at Rainier, Oregon.

The power from the thermal plant is delivered to the BPA regional transmission grid. Here, Bonneville "acquires" the power and enters a credit in its ledger to the account of the participating utility. The 100 megawatts from CNP, at 20 mills/kwh, amounts to $17.5 million over a year; this is credited to Sasquatch.

At the delivery end of the regional grid, BPA transfers power to Sasquatch. But it provides not only the 100 megawatts from CNP, but power from the Columbia River dams as well--300 megawatts in the example. At the close of each accounting period BPA presents the public utility with a bill for the power it has received. The credit for the thermal power is deducted from the bill for the total amount of power received by Sasquatch--400 megawatts priced at 7 mills/kwh in the example. Thus, Sasquatch pays two bills, one to Bonneville and one to CNP. The wholesale power rate (7 mills/kwh) is a "melded" price, intermediate between the cost of thermal (20 mills/kwh) and the cost of producing federal hydro (typically less than 2 mills/kwh).

Two legal principles are at work here. First, no money has flowed through BPA for the "net-billed" power: a credit was entered, to the utility's favor, and a debit was billed against its account. Therefore, BPA has served "only" as a marketing agent: it has not bought and sold *for cash* the power from the thermal plant. Its actions are thus in accord with the Bonne-

Figure 13. *Net Billing in the Hydro-Thermal Power Program.*

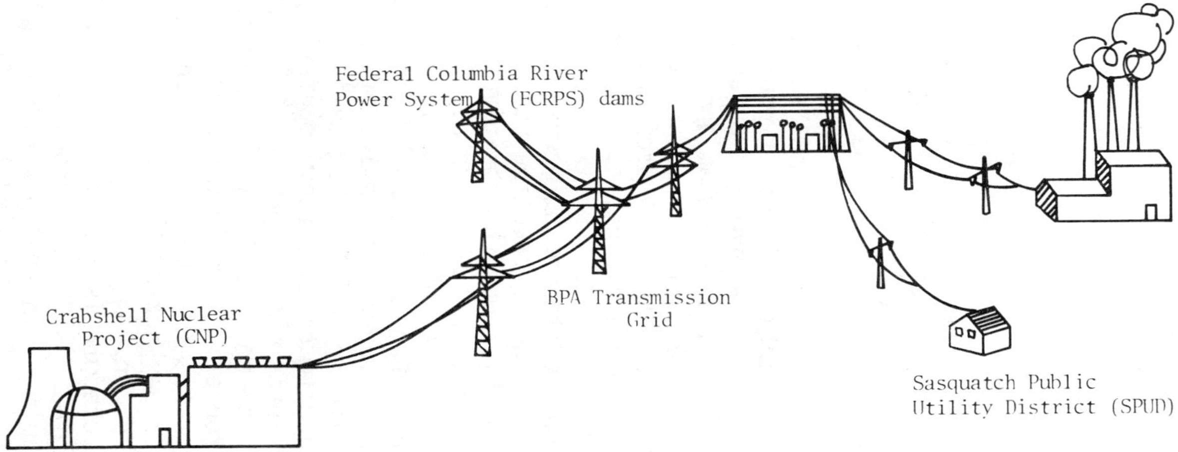

BPA Power Account

	received from CNP	delivered to SPUD
	100 MW	400 MW
Power Flows:	CNP to SPUD, via BPA	100 MW
	FCRPS to SPUD, via BPA	300 MW

BPA Bill

Credit:	for power received	@ 20 mills/kwh	$17.5 million
Debit:	for power delivered	@ 7 mills/kwh	24.5 million
Net owed to BPA			$ 7 million

Dollar Flows:	SPUD to CNP	$17.5 million
	SPUD to BPA	$ 7 million

ville Project Act, which gives BPA authority to distribute electricity on a regional basis, but not to purchase non-federal power for resale. Second--following the language used in hydro contracts--Bonneville is legally committed to deliver power to its preference customers on a firm basis.[31] This means in particular that BPA is obliged to supply the total power--400 megawatts for Sasquatch PUD--even if the thermal plant should produce no output at all. Thus, the credits for thermal power are entered in the ledger whether or not the thermally generated electricity actually flows into the regional grid. In this manner, the public utility participant is insured by BPA against the failure of the thermal plant. If a thermal project turned out no power at all--what utilities call a "dry hole"--BPA wholesale power prices would be raised throughout the region to cover the costs. Accordingly, the costs would be shared among all buyers of the federal hydro resource, not just participating utilities, for whom a dry hole could prove catastrophic. Net billing was an offer too good to refuse. Of BPA's more than 100 preference customers, only the City of Tacoma did not take advantage of net billing.

Net billing is a limited financing mechanism, since the thermal power credit cannot exceed the public utility's total bill. The net bill must be positive. In practice, the thermal credit is limited to about 85 percent of the total bill, so that BPA can derive enough revenues to stay solvent. BPA is responsible to the U.S. Treasury to collect revenues from sales of federal hydro sufficient to retire the debts incurred in building the hydro system, and to cover its own operating costs. So long as the rates charged public utilities are high enough that these "payback" payments may be made to the Treasury, Bonneville is in the clear. In recent years BPA power rates have been raised in order to maintain revenues for payback as the net-billed portion of hydropower has grown.

This accounting procedure is legally intricate but economically straightforward. BPA purchases non-federal power and shares the risks of thermal power development by putting Bonneville's guarantee behind the delivery of power to preference customers. With net billing public utilities could readily obtain borrowed funds for plant construction. Thus net billing transforms the public preference policy into financial leverage. The preference policy creates property rights, rights to federal power, for the public utilities. These rights are realizable as income in the form of annual deliveries of federal hydro. The net-billing scheme turns income into equity, equity available for capital investment.

When it was originally devised, net billing had to be submitted for approval to the Appropriations Committee of Congress. Even though net billing stayed within the marketing authority granted by the Bonneville Project Act, the procedure entailed financial obligations which needed congressional review.[32] Like other executive branch proposals, net billing had first to be scrutinized by the White House Office of Management and Budget. OMB objected on the grounds that it seemed too much like a guarantee to bail out public utilities. As Bonneville power manager Goldhammer remembered the criticism:

> ...what do you do if these plants don't get into operation? Who pays for them? Are you going to come back and ask for an appropriation to bail out the costs for a plant that is not operating for any reason?...The question kept coming up, why can't the utilities in the Northwest finance and go ahead with these plants?... Why should Bonneville be underwriting and spreading out the cost of a dry hole?[33]

In response, the congressional General Accounting Office was asked for an advisory opinion on the legality and fiscal prudence of net billing; GAO supplied a brief statement, indicating that the procedure was permissible. Remarkably, neither the executive branch OMB nor the legislators' GAO seems to have consulted their experts in the field. "I would have raised a number of questions" about the open-ended financial commitment, one person involved told us.[34] As the costs of thermal plants skyrocketed, the unasked questions became pregnant with meaning: "No one anticipated how hard it would be to build these turkeys."[35]

An intrinsic part of net billing was a price structure that averaged low-cost federal hydro with high-cost thermal energy. From the standpoint of economic theory, cost averaging is misleading. In order to achieve economic efficiency, consumers should make decisions about purchases on the basis of marginal or incremental cost. That is, the price charged for a commodity should be the cost of making a new, or incremental, unit of that commodity. (This economist's ideal is rarely achieved in practice, however--and practically never in the electric power market.)[36] In an expanding market for electricity, consumers should pay a price set by thermal generation, the latest unit added to the power supply. The net-billed average price does not do this, and in that sense it underprices electricity. Something that is charged too low a price is apt to be bought in overlarge quantities, and critics of HTPP have alleged that net billing unduly promotes growth in power use and hinders conservation efforts.

Moreover, net billing, in combination with the postage-stamp pricing policy, produces inequities among utilities, once again favoring those preference customers whose load is growing most rapidly. The postage-stamp philosophy dictates uniform rates to all customers within a class, so that all public utility requirements contracts are set according to a uniform rate schedule. But as pointed out above, the wholesale power rates establish limits to the amounts of financing that can be developed under net billing. Consequently, as thermal plant costs rise, pressure mounts to raise wholesale rates for firm power. This expands the fiscal ambit of net billing, but it simultaneously raises everyone's power costs, including those of utilities which are growing slowly and thus contributing less to the need for high-cost thermal power. Even Tacoma, which did not participate in net billing at all, must pay the higher wholesale rates necessitated by HTPP-sponsored thermal plants.

The net billing procedure is a monument to the politics of public preference. The judgment that public preference was too deeply entrenched to be repealed led to a convoluted institutional arrangement with managerial complexities and, ultimately, economic inefficiences. But it brought the public utilities of the region into the Hydro-Thermal Program.

The Ten-Year Program

In 1969 the Joint Power Planning Council announced a ten-year, $7 billion program of cooperative investment involving 109 utilities and BPA. Seven thermal power plants would be built, at an estimated cost of $2 billion. The federal government would build additional hydro facilities to enhance the peak power capabilities of the Columbia River system, at a cost of $1.3 billion. And federal, public, and investor-owned utilities would build additional transmission facilities to serve growing customer needs, to the tune of over $3.7 billion (Table 7).

The JPPC, chaired by BPA administrator H.R. Richmond and led largely by Bonneville staff, had succeeded in laying out a program of region-wide cooperation. The central objective was orderly growth in the regional power supply. Technologically, the hydro-thermal concept would be realized, if in a provisional way and on a limited scale. Financially, significant economies of scale would be captured in the construction of large thermal plants, which seemed in the 1960s to be proportionally less expensive than smaller ones. In organization, the Hydro-Thermal Program placed Bonneville in the crucial coordinating role. Its responsibility to market federal hydro and to operate the regional transmission grid would be expanded in HTPP to encompass the management of net billing for the public utilities. Net billing was the major incentive BPA provided to proceed with thermal power. It distributed the financial risks attending any single thermal plant over all the Northwest's public utility customers. A second type of insurance was embodied in BPA's provision of regional reserves. These were to meet unexpected surges in demand and to compensate for unplanned outages and construction delays at facilities covered by net billing.

HTPP marked a political milestone as well: the achievement of a regional orientation in power planning. In HTPP both the divisions of the public power fight and the suspicions of the Republican-enforced partnership between IOUs and public power seemed to be transcended. The sharing of regional benefits--no less real for being measured by the parochial yardstick of the utilities--carried forward the precedent of the N-Reactor. But unlike the N-Reactor, the federal support for regional growth was now institutionalized in Bonneville, freed from dependence on Northwest congressmen's flair for legislative strategy. The design of HTPP, moreover, brilliantly harnessed the public preference policy. A major obstacle to public utility participation in the hydro-thermal system--the financial characteristics arising from public power's historical dependence on federal hydro--was converted through net billing into the primary means for participation. The Hydro-Thermal Program achieved the technical, economic, and--if only temporarily--political integration of the regional power system.

HTPP was to begin in 1971, and last until 1981. Within two years it had to be revised fundamentally.

The weak point in the program, perhaps unsurprisingly, was net billing. HTPP also ran afoul of a number of external complications: environmental opposition, the energy crisis in petroleum, and the related economic recession of 1974-75. These external forces and the program proposal known as Hydro-Thermal, Phase II, will be described later. Here it is worth noting the two difficulties that beset net billing by 1973.

The major difficulty was skyrocketing cost. The cost of building thermal power plants has mounted much faster in the 1970s than had been expected on the basis of experience in the 1960s.[37] The experience of WPPSS, while apparently extreme, illustrates the onrush of economic peril that threatens to engulf the utility industry.[38] When it was listed as one of the seven original plants in the Hydro-Thermal Program, WPPSS Nuclear Project No. 2 (WNP 2), sited in Hanford, Washington, was scheduled to cost $352 million (Table 8), and it was expected to be in operation by September 1977. WNP 2 is already more than three years behind schedule; it will cost about $1.82 billion. WNP 1, another nuclear project at Hanford, was originally announced at a price tag of $632 million; along with delays, WNP 1 will come in at more than $2.3 billion. Building nuclear plants is not for the fainthearted: the Supply System's cost estimates for its five plants under construction have proved in the aggregate to be at least $7.75 billion too low, and their construction schedules have slipped more than a decade. These massive shortfalls are just beginning to be felt in the wholesale rate structure, since repayment for the costs of the plants does not ordinarily commence until the plants begin operation. But the delays have now accumulated to such a point that interest payments on revenue bonds, scheduled originally according to optimistic completion dates, have begun to fall due: BPA is now paying $6 million per month on WNP 2.[39] BPA has announced that its wholesale rates will increase by about 90 percent in early 1980; it is only the beginning.

Net billing was limited in the amount of capital that could be harnessed on behalf of the public utilities. The cash flow--in effect, the mortgage payments for capital investment in thermal plants--could never exceed roughly 85 percent of the dollar value of BPA's billings to its preference customers. With rising costs of construction, the capital needed to build HTPP's thermal plants was rapidly exceeding the supply that could be backed by net billing. Because hydro used in net billing was sold via long-term contracts, its sales price could not be easily adjusted; the Bonneville Project Act provides that rates may be adjusted no more than once every five years.[40] Early in 1973, BPA estimated that by 1981 the revenues derived from the sale of federal hydro to preference customers would amount to $145 million. But the annual cost of the thermal plants built under HTPP by 1981 would have mounted to $135 million, over the net-billing limit of 85 percent.[41]

The Hydro-Thermal Program hit a second snag in 1972, when a new federal tax ruling sharply limited the possibilities of cooperative plant investments between public and investor-owned utilities. The Revenue and Expenditure Act of 1968 mandated changes in the tax exemptions allowed joint public-private ventures to reduce opportunities for private enterprises to use public authorities to shelter their earnings from federal taxation. By an administrative regulation effective in August 1972, the Internal Revenue Service set a limit of 25 percent on the amount of power that may be sold to a "non-exempt" buyer from a publicly owned power plant.[42] While this did not foreclose cooperation in planning, it further rattled the already unsteady framework of HTPP.

In any case, net billing was abandoned for projects begun after 1973, though it has been retained for plants whose financing was already underway. Once the linchpin of federally backed financing was removed, the regional power program gradually came apart, as internal tensions--no longer managed through Bonneville's control of net billing--reinforced the centrifugal effects of external events and opposition.

SCALE, PLANNING, AND TECHNOCRACY

The difficulties encountered by the Hydro-Thermal Program should not obscure the magnitude of the achievement it represents. However temporarily, Bonneville's leadership crystallized a region-wide utility program and transcended historic disputes surrounding public power. That this could be done sets an important precedent for the future. That it was done in the name of technological and economic forces which claimed to supersede political rivalry suggests a diagnosis of the present impasse. The surprisingly apolitical tenor of HTPP reflects the high technical, fiscal, and organizational complexity of the hydro-thermal concept as it took institutional form. BPA, as one congressional aide commented, "is a cross between a private enterprise organization and a governmental organization."[43] Nowhere was this more the case than in HTPP.

But the complexity of its institutional form made HTPP both rigid and fragile. Once it wobbled financially, the program became susceptible to criticisms, such as the claim of economic inefficiency in the melded price structure. These complaints could not be easily accommodated because the web of interdependence was hard to re-spin once torn. Like other gargantuan technological undertakings, the complexity of HTPP made it peculiarly vulnerable to the procedural attacks of critics, especially environmentalists versed in the tactics of litigation. As HTPP unraveled, the scramble for low-cost federal hydro reopened the half-healed wounds of the public power debate.

The brittle institutional structure of large, complex technical systems is cast in sharp relief by comparing the hydro-thermal concept with its hydroelectric predecessor, using the attributes in Table 9. First,

Table 9. *Six Attributes of Large-Scale Scientific Technology.*

(1) Long lead times
(2) Capital intensity high
(3) Inflexible commitments of resources
(4) Highly trained manpower
(5) Centralized organizational control
(6) Planning becomes paramount

Source: Adapted from John Kenneth Galbraith, *The New Industrial State* (New York: New American Library, 1971 rev. ed.) Chap. 2.

the incorporation of thermal technology substantially increases the lead time, the interval separating a decision to build a generating facility from its actual operation as part of the regional power supply. Although dams too have been delayed in recent years, often by external opposition, thermal plant lead times are inherent in the complexity of the technology. The precise, large-scale welding needed to build a nuclear plant, for instance, has no analogue in a hydro project.

Second, the capital needed for thermal power development is proportionally high. Hydro generation is likewise capital intensive, but the Northwest hydro system was built during a period of low construction costs, with financing at low interest rates; federal backing cut borrowing costs even further. There were additional advantages in the multiple benefits of taming the river, which made it possible to apportion the costs of federal dams among power rates, irrigation, navigation, and flood control. The lower capital costs of multiple-purpose dams, together with the fiscal reach of national government, made it possible for the federal government to finance Columbia River hydro on its own; thermal power, as we have seen, requires cooperative financing.

Third, the hydro-thermal concept requires tightly coordinated, inflexible commitments of economic and natural resources. But its most dramatic manifestation is the impact on the Columbia River, as it is drawn increasingly into producing peak load power for the hydro-thermal system. Some of the multiple uses on which the hydroelectric investment was built are constrained or simply made infeasible under the full development of the hydro-thermal regime--though the extent of these use conflicts is even now not clearly established.

Besides the trend to capital intensity, HTPP in two further respects extends patterns of change already visible in the hydro development. The need for specialized, highly trained technicians is far greater in the hydro-thermal scheme, and the need for coordinated management and centralized organizational control is enhanced. The upper Columbia basin hydro development program launched in 1964 had already highlighted the need for coordinated operations and the expertise to carry them out. The extension of coordination into investment under net billing enlarged the managerial burden of--and authority embodied in--Bonneville's marketing responsibilities.

The sixth and paramount difference between hydro and hydro-thermal is the emergence of planning and investment as the central focus of HTPP. During construction of the hydro system institutional arrangements could be made one project at a time. The Columbia River power agency is still named for the earliest major--but by no means pre-eminent--dam. The focus of dispute in the hydroelectric period was ownership and its concomitant operating practices and philosophies. The cooperative Hydro-Thermal Program honors ownership formally in the legalisms of net billing, while purposefully playing off different kinds of property rights to tease equity capital for investment from the white water in the federal turbines. And the five other attributes of technical complexity lead inexorably to ever more elaborate preparation for the technological feat itself. Thus do planning and investment rise to the fore.

As the demand for electricity expands to the limits of the low-cost hydro resource, the era of cheap power --and the program of social development which it was intended to foster--is coming to an end. The welding of the Columbia hydro projects into a regional generating program requires cooperation, not the conflicts of public power. HTPP promised to keep the benefits of cheap, abundant power for as long as possible: economy would be achieved via large-scale coordination and abundance would be won from cooperative operation of river and thermal power stations. But for this, planning was indispensable. Accordingly, planning became the new focus of debate, a debate characterized by the passionate abstractions of voices vying for control of the not-quite-tangible future.

These six differences between the hydro and hydrothermal periods are the institutional hallmarks identified in Galbraith's analysis of *The New Industrial State*.[44] By these criteria HTPP may be seen as the product of a "mature corporation." The mature corporation differs from the entrepreneurial firm of economic theory --and free enterprise ideology--chiefly in its size and stability. Unlike its smaller forebears, the mature corporation can operate with a significant degree of independence from market forces; Galbraith argues, indeed, that it can influence its own markets and sometimes control them. In transcending the market the mature corporation dispenses with the single entrepreneur. The capitalist, according to Galbraith, is displaced by the "technostructure," a somewhat fuzzily defined agglomeration of technologists, managers, and financial analysts whose expertise, cumulatively and collectively, guides the large corporation.[45]

But of course, the Hydro-Thermal Program has no mature corporation behind it. There is no firm at all, only a collection of more than 100 organizations of varying pedigree, ownership, and economic interest. Unlike Galbraith's super-corporations, the institutional structure of HTPP has been notably unstable, and the regional power program continues to be riven by tension and competition. What is important about the mature corporation model is not its literal applicability, but its utility as an ideal benchmark: the goal toward which much of the maneuvering by leading actors in the regional power planning arena is directed.[46]

The need for technical coordination has, of course, long been recognized. Currently, more than 100 of the region's publicly owned utility systems engage in coordinated technical study, forecasting, and overall supply planning through the Public Power Council (PPC). Publicly owned utilities which own generating resources coordinate among themselves. Through the Public Generating Power Pool (PGPP) the utilities optimize the use of non-firm resources by taking advantage of their load diversities.[47] The Washington Public Power Supply System and the Pacific Northwest Generating Company, formed in 1975, also provide coordinated facilities planning and financing for publicly owned utility systems. In 1973, the IOU systems began to plan cooperatively via the Intercompany Pool.[48] (See Table 10).

The most visible utility group is the Pacific Northwest Utilities Conference Committee (PNUCC). Formed in 1946, PNUCC is composed of nineteen public and four investor-owned utility members. Its claim to representing the utility industry of the region rests on a policy committee, formed with representation from 130 utilities and direct service industries. PNUCC is the only region-scale planning organization which includes all of BPA's wholesale customers.[49] The PNUCC publishes the West Group Forecast, a compilation of individual utility forecasts which serves as a planning guide for the region's utilities. The PNUCC also provides technical analysis of regional electrical issues through task forces organized from utility staffs. Until it took a leading role in drafting regional power legislation in 1977, however, the PNUCC was not publicly active in political discussion.

Of the plethora of utility coordinating bodies, it has been Bonneville which has traditionally been accorded primacy. Had HTPP worked as originally planned, Bonneville would have assumed the role in the hydrothermal era that it had through much of the hydroelectric era, that of technostructural manager and political leader. With the unraveling of HTPP, technostructural hegemony can no longer be exercised, though there are contenders for the top spot, notably BPA and WPPSS.

The utility and BPA managers of the Hydro-Thermal Program resemble Galbraith's technostructure in a number of ways. The most salient features are the dominance of growth as a goal and of technical virtuosity as a style. Galbraith argues that the managers of the mature corporation benefit only secondarily from profits, which accrue to stockholders, only a small fraction of whom are managers. Hence the technostructure emphasizes growth, which confers primary benefits on itself through enhancement of organizational power and size.[50]

Table 10. *Membership of Northwest Utilities and Agencies in Coordinating Groups.*

	Organization or Agreement						
Investor-owned Utilities	Northwest Power Pool (NWPP)	Pacific Northwest Coordination Agreement	Western Systems Coordinating Council (WSCC)	Pacific Northwest Utilities Conference Committee (PNUCC)	Mid-Columbia Hourly Coordination Agreements	Columbia River Water Management Group	Intercompany Pool (ICP)
Colockum Transmission Co., Inc.		X			X		
Idaho Power Company	X		X				X
Pacific Power & Light Company	X	X	X	X	X		X
Portland General Electric Company	X	X	X	X	X		X
Puget Sound Power & Light Company	X	X	X	X	X		X
The Montana Power Company	X	X	X				X
The Washington Water Power Company	X	X	X	X	X		X
Utah Power and Light Company	X		X				X
West Kootenay Power & Light Co., Ltd.	X		X				
Public Agencies							
British Columbia Hydro & Power Authority	X		X				
Eugene Water and Electric Board	X	X	X	X			
PUD No. 1 of Chelan County, Wash.	X	X	X	X	X		
PUD No. 1 of Cowlitz County, Wash.		X	X	X	X		
PUD No. 1 of Douglas County, Wash.	X	X	X	X	X		
PUD No. 1 of Pend Oreille County, Wash.		X		X			
PUD No. 2 of Grant County, Wash.	X	X	X	X	X		
Seattle City Light	X	X	X	X	X		
Tacoma City Light	X	X	X	X	X		
Washington Public Power Supply System	X			X			
Federal Agencies							
Bonneville Power Administration	X	X 1/	X		X	X	
Bureau of Reclamation	X	X 2/	X			X	
Corps of Engineers, North Pacific Div.	X	X 1/	X		X	X	
Others							
26 Western utilities			X				
9 Northwest PUDs; 3 Northwest Municipal Systems				X			
8 Federal agencies, including EPA, FPC, USGS, USFS, BSFW, NMFS, SCS, NWS: 6 state agencies						X	

1/ BPA, acting for the Department of the Interior, and the Corps of Engineers, acting for the Department of the Army, signed the Agreement as respresentatives of the United States of America and the United States Entity.

2/ The Bureau of Reclamation, although not a signator of the Coordination Agreement, is represented on the Coordination Contract Committee.

Source: BPA, Division of Power Management, Branch of Power Supply.

Utilities are constrained in their profitmaking in any event because the IOUs' profits are regulated and public agencies' "earnings" are redistributed via lowered rates. IOU stockholders are remote from the managerial scene, and public system ratepayers are quiescent so long as rates are stable; in both cases, growth is a readily embraced goal. Now that rates are rising, growth--which is largely responsible for higher costs--has come under attack.

HTPP is replete with technological, economic, and legal innovations devoted to optimizing the benefits of growth and to fostering utility cooperation. This virtuosity was purchased at the price of ideological clarity, as the New Deal preference for public agencies evolved in HTPP into the fiscal minuet of net billing. The retreat from ideology was fundamental to the choice to pursue the hydro-thermal concept; both publics and IOUs had to participate if hydro-thermal was to work. But Bonneville's devotion to public service, visible and controversial at its founding, was diluted into mere growthmanship as demand mounted beyond the capabilities of the hydro system. The Hydro-Thermal Program was too complex for ready public interpretation, and public debate was in any case discouraged by Bonneville power managers who were anxious to nurture the healing of the public-private rift. Yet growth proved to be an inadequate substitute for the populism of public power and its promise of social development, especially once costs began to rise.

Although BPA retreated from public leadership, the agency nonetheless remains the core of the regional electric power technostructure. Its staff, now numbering more than 3,000, provides information and analytic assistance to the planning efforts of the region's utilities. In particular, BPA develops load forecasts for nearly 90 percent of the public systems, mainly the smaller ones serving rural areas. And Bonneville analysis contributes also to the compilation and synthesizing of the West Group Forecast of the PNUCC, the region's most comprehensive--hence influential--projection of future demands and resources.[51] More broadly, the administration of HTPP demanded precise orchestration of wholesale power production and planning by an elite staff, mostly in Bonneville, together with the utilities, DSIs, and the sponsors of thermal power stations. (It is this team which crafted the PNUCC Bill in 1977, a bill that would have authorized those same managers to plan and invest in power supplies for the Northwest.) Yet differences of organizational allegiance and political ideology have divided the regional technostructure, while congressional legislation seemed in its very wording to rekindle the debate over public power.

The divisions within the technostructure may be suggested by the rising and controversial influence of two entities contesting--in quite different ways--Bonneville's leadership: the Oregon Department of Energy (DOE) and the Washington Public Power Supply System. Founded in 1975, the Oregon DOE is the only state energy agency with sufficient analytic depth to develop its own electric power demand forecast.[52] Moreover, it has challenged the forecasting and planning methodologies adopted by the utilities, for example in the licensing proceedings for the Pebble Springs nuclear plant.[53] Though Washington and Idaho have formed energy offices, they are little more than information gathering and dispensing bodies; they make no significant impact on long-range electrical planning issues. Montana's energy office has only recently become involved in the Northwest controversy, though Montana state officials did play an active role in congressional legislation in 1979. State siting councils have varied in the degree to which they scrutinize need for power when a plant is proposed. As a result, utilities retain essentially all planning responsibilities traditional to their role, even though political acquiescence in utility expansion has disappeared. The utilities find themselves, accordingly, in the anomalous position of exercising responsibilities whose legitimacy has been dramatically eroded.[54]

Oregon DOE's challenge is analytic; WPPSS's is technical and economic. Though chartered in the state of Washington, WPPSS markets participant shares in its new plants throughout the region. In this way, it is providing a powerful central planning force for public utility systems, which were fragmented in the past by their variety in size, geographical distribution, political and economic bases, and demand and generation characteristics. The traditional reliance of many smaller utilities on BPA for assistance and policy guidance may accordingly be changing. The growing influence of WPPSS is reflected in the growth of the organization as thermal power increases its role. By fiscal year 1977, WPPSS staff had grown to 714, up from fewer than 100 in four years. Its five nuclear plants now under construction will cost an estimated $11.75 billion.[55] WPPSS is also currently involved in over $100 million worth of energy projects to plan for additional generation and to acquire more than ten million pounds of proven uranium reserves by the mid-1980s.[56] A major objective of WPPSS has been to develop its own planning and engineering staff so that fewer consulting contracts would have to be let. Thus, WPPSS is evolving toward the highly specialized staff structure of BPA.

The future relationship of BPA and WPPSS is uncertain. The legislative context in which BPA operates limits its role in providing generation. With the shift in emphasis from hydro to thermal technology, however, WPPSS may unite the public utilities under the umbrella of the research, financing and operational services it can provide. Either BPA or WPPSS could provide, individually or collaboratively, the central element in institutional redesign, by virtue of the regional orientation natural to its technology.

In the short term, the large cost increases in WPPSS nuclear projects, three of them funded through net billing, have brought the Supply System into confrontation with Bonneville.[57] A number of audits of the performance of WPPSS and its contractors have been carried out, including one by the Office of Audit and Investigation of the U.S. Department of the Interior, BPA's federal home before it was reorganized into the new Department of Energy. The boundary between Supply System managerial independence and Bonneville oversight responsibility has been a hotly contested one. This appears to be a contributing reason for WPPSS's attempt to put itself forward, early in 1978, as the effort to enact regional power legislation was collapsing, as the champion of public power in the Northwest. (Of this, more in Chapter V). But coordinated development of the hydro and thermal technologies will remain problematic so long as BPA and WPPSS stay on bad terms.

There are a number of reasons why HTPP faltered. Here, we point out a larger, elusive pattern, one not at all plain in the contention over who should get the benefits of cheap federal hydro: the political and economic character of the technology itself. The requirements of region-scale thermal power development are ambitious, and, as Galbraith argued,

> the firm is required, with increasingly technical...processes, increasing capital, a lengthened gestation period and an increasingly large and complex organization, to control or seek to control the social environment in which it functions....It must plan not only its own operations; it must also, to the extent possible, plan the behavior of people and the state as these affect it. This is a matter not of ambition but of necessity.[58]

The juggernaut tendency of an activity which requires planning "the behavior of people and the state" is discomfiting, if not frightening. That is part of the reason why HTPP and its successor proposals have encountered political opposition from environmentalists and public power advocates.[59] But the need to transcend uncertainties, both those of the marketplace and those of politics, seems disturbingly characteristic of the large-scale technologies that are the *metier* of the mature corporation:

> It is open to every free-born man to dislike this accommodation. But he must direct his attack to the cause. He must not ask that... nuclear power plants...be produced by firms that are subject to unmanaged demand. He must ask instead that they not be produced.[60]

If this problem, deriving as it does from the character of the technology and not from how it is managed, is real, there are substantial limits on how stable *any* regional power plan can be.

One of the tantalizing aspects of Galbraith's analysis is his observation that universities have become a major source of criticism of the new industrial state.[61] The criticism is perversely symbiotic, since universities also train those who become the technostructure. Thus, the development of scientific knowledge from which the new industrial state springs nourishes challenges to its technostructural mandarins. As with the technostructure's own claims to authority, much of its critics' opposition is grounded in scientific analysis. This ambivalent contribution of knowledge to politics inspires both the hydro-thermal dream and the determination of its environmentalist opponents.

CHAPTER IV
The Broadening Agenda

> The main trouble with the world today is
> that the future is not what it used to be.[1]
>
> --Paul Valery

After four decades of growth and frequent conflict, electric power development in the Pacific Northwest seemed in the early 1970s to be moving toward cooperation and consensus. Cooperation among privately owned utilities, public power agencies, and the federal government was overcoming the divisions inherited from the New Deal. Part of the price of cooperation was a deflection of the purposes that had surrounded the majesty of hydropower dams with the democratic aura of rural development. But growth in electrical demand seemed so insistent: to the utilities growth was at once the beacon of progress and the lash of necessity.

Growth became deeply problematic in two senses. Environmentally, growth seemed to promise further industrialization, threatening a part of the country that still prided itself on a certain refined rusticity, that hungered, as only wealthy societies can, after "quality of life." Economically, growth turned out to be unexpectedly, suddenly expensive, especially for a regional economy used to the extraordinarily low costs of federally developed hydropower.

Environmental quality and mounting economic costs each had protean qualities: each appeared and reappeared as concerns, complaints, and challenges to the technocratic world view of the utilities. In the end, these concerns could be neither deflected nor opposed, and regional power planning collapsed. In this chapter we shall indicate those burdens that traditional power planning could not bear--although the social currents we shall discuss run far deeper and broader than either the Pacific Northwest or energy policy.

The high cost of growth gave urgency to cost sharing decisions. These decisions, especially the sharing of federal hydropower, have become the principal source of political disarray. But if economics is now the issue, environmental criticism gave the economic pressure time to build. For in the mid-'70s, as the Bonneville Power Administration and the utilities sought to make arrangements to go beyond the financing capabilities of net billing, environmental intervention thwarted federal leadership. In 1975, Bonneville administrator Don Hodel lashed out at those whom he labeled "prophets of shortage":

> Their tactics are those of confrontation rather than constructive dialogue. Their weapons are the injunction, the restraining order, and the challenge to the environmental impact statement. Their theater of operations is the courtroom, where they are always the plaintiff, never the defender...the anti-producers do it week in and week out, seldom winning a case but satisfied to keep progress at bay with their legal maneuvers.[2]

Without this delay, a regional power program might have been launched, and long-run commitments made, before the cost implications of thermal power hit consumers. The original Hydro-Thermal Program described in Chapter III was intended to do just this: it provided for a smooth transition without disturbing consumers' assumptions that things were fine in electric power.

Yet as Hodel admitted in his 1975 speech, "I cannot attribute all of these [problems] to environmental pressures. Erratic federal funding, labor disputes and shortages, technical problems--all conspire to upset schedules and impede new generation from coming on line."[3] In light of the divisions within the electricity utility industry which have become public since 1977, Hodel's attack seems in retrospect an attempt to identify a scapegoat--one not too dangerous to criticize. If the regional power debate leads to a program superior to the utilities' hydro-thermal plans, the Northwest will have cause to thank its prophets of anticipated costs. That remains a troublesomely large "if."

Until federal legislation was introduced by Senator Henry Jackson in 1977, environmentalists took the lead in highlighting regional electric energy issues, bringing them before governmental forums and consumer groups. But well before economics reignited the banked fires of the public power debate, the environmentalist agenda had narrowed dramatically. Protection of the environment itself became an indirect, sometimes indistinct goal. Instead, energy conservation, renewable resources, and accountability to the public--mainly via state government--came to the fore. This focusing of energies reflected an attempt by environmental leaders to develop a political agenda appropriate to the scale of HTPP, as well as to forge alliances with others outside the traditional utility-BPA-congressional establishment. As we shall discuss in Chapter V, the outsiders' insistence on being included in power planning is likely to be accommodated to some degree. It remains to be seen how well environmental quality will be served by that accommodation.

ENVIRONMENTAL COSTS

Continued expansion of electric power generation will bring new and larger environmental stresses to the Pacific Northwest. In addition to impacts from thermal plants, there will be further changes in river habitats as hydro facilities are increasingly used to meet peak demand. The BPA Role EIS summarizes some troubling projections.

Hydropower and Peaking

Fish. Existing development has already fundamentally altered the Columbia River fisheries. The building of Grand Coulee Dam cut off the upper Columbia from migrating fish, among which the majestic salmon is symbolically and economically most important. Salmon runs that once ranged 1,200 miles along the Columbia are now stopped by the impenetrable walls of Chief Joseph Dam on the Columbia and Hells Canyon Dam on the Snake, each 600 miles in from the Pacific.[4]

The dams, by slowing the flow of water, appear to slow the downstream migration of young salmon.[5] Remarkably, when fish pass through the power turbines they often survive. But the smaller juvenile fish are stunned, and many fall prey to squawfish which flourish in the stilling ponds below the Columbia River dams. The mortality of juvenile fish passing through turbines has been estimated to range between 5 and 30 percent per dam.[6]

In addition, the spilling of water over the dams leads to nitrogen supersaturation; the water in its long fall entrains extraordinarily large quantities of nitrogen. Fish downstream from the dams absorb this nitrogen into their bloodstreams in respiration. In a phenomenon analogous to "the bends" experienced by divers, when fish come near the surface of the water the nitrogen can come out of solution in the blood, injuring or killing them. This "gas bubble disease" was only recognized in 1965.[7]

Flood Control. Beyond the problems that have become apparent under present operating practices, there is substantial uncertainty about the future. The broader dimensions of competition between the non-power uses of the Columbia River system and the hydro-thermal program are sufficiently complex to defy precise estimates.[8] But the central feature of the hydro-thermal approach is clear-cut: the seasonal variation of flow, the key determinant of river management in the past, will give way gradually to year-round husbanding of the water resource.

The upstream storage reservoirs--those above and including Grand Coulee Dam--have formed the basis for providing hydropower during the low natural flows of winter, together with storage to hold back springtime floods. The hydro-thermal approach calls for the conversion of Grand Coulee into a peaking plant, to meet the daily and seasonal fluctuations of demand (see Figure 14). The economic logic of this conversion leads

Figure 14. *Estimated Firm Generation, Pacific Northwest (Typical January Mondays).*

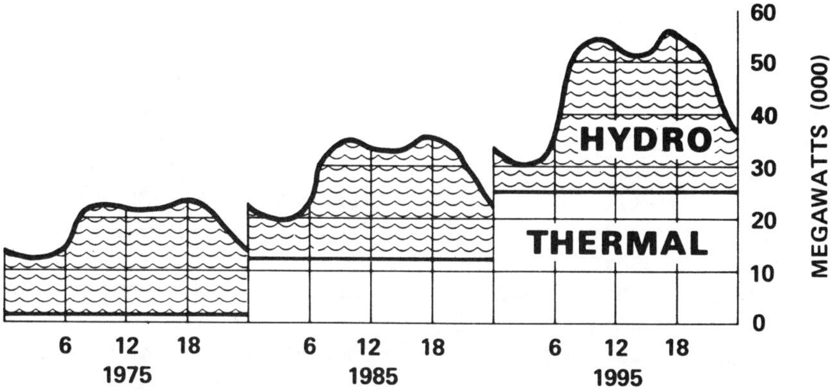

Source: Bonneville Power Administration, *The Electric Energy Picture in the Pacific Northwest* (Portland: BPA, 1976).

to the conclusion that the Grand Coulee Reservoir, Lake Franklin D. Roosevelt, should be kept as full as possible throughout the year. This way, the "head," or distance through which the water falls in the turbines, can be maximized and with it the peaking capability of the hydro plant.

Traditionally, however, Lake Roosevelt has been operated for flood storage: during winter and early spring, the lake has gradually been drawn down to make room for spring floodwaters, while also providing water for power generation during the low-flow season. As the region shifts toward a hydro-thermal configuration, the demands of winter power generation and flood control will become competitive rather than complementary. Reconciliation of these objectives is possible, but the cost of pursuing both will rise.[9]

Dams downstream from Grand Coulee further complicate the trade-off of power and flood control. For both physical and political reasons the downstream dams have no significant storage capacity. The partnership policy, by withdrawing federal resources, led inadvertently to a pattern of development emphasizing hydropower, which produces revenues, at the expense of other uses such as flood control or navigation, whose economic benefits are less readily collected. Dams built under "partnership" arrangements cluster on the mid-Columbia reaches of the river below Chief Joseph.

With no history of stewardship for flood control or other non-power uses, the mid-Columbia utilities that own "partnership" dams have shown little willingness to pursue anything but the maximization of power revenues. Yet the profile of water releases upstream, especially from Grand Coulee, affects the water available at mid-Columbia dams for peaking--the most valuable segment of hydro. The move toward hydro-thermal operation promises little of benefit for fish or flood control.

Daily fluctuations. The changes in seasonal policies under a hydro-thermal concept are matched by shifts in daily operations. As the demand for power grows, the relationship between average and peak power demand is expected to be maintained.[10] The daily peak will thus grow, and in the hydro-thermal approach these variations are handled by the hydro system. In 1975 the daily cycle of electricity demand caused hydro output to vary by a factor of two between its early-evening peak and the late-night minimum. By 1995 the ratio is projected to be a factor of six (Figure 14). As early as 1985 the level of the tailwaters at the foot of Grand Coulee Dam is conservatively estimated to fluctuate by fifteen feet.[11] (Today, the maximum variation is nine feet.) These changes in level will be damped by about half five miles downstream, but will then retain that fluctuation until reaching Chief Joseph Dam, fifty-one miles away.[12]

The ecological impacts of these changes are uncertain. There is evidence that the fluctuations in reservoir levels can inhibit the upstream passage of mature fish. In 1974 a retrospective study of fish counts at Bonneville, John Day, and The Dalles dams demonstrated that peaking operations were correlated with decreased counts of migratory fish. A 1975 study at The Dalles reconfirmed this correlation. Subsequent detailed studies undertaken by the Army Corps of Engineers and the National Marine Fisheries Service could not, however, demonstrate how this correlation influences fish behavior.[13] This led Bonneville to conclude that "it is impossible to predict with confidence the effects of the developing system on adult salmonids,"[14] although there is no indication that the effects will enhance fish populations.

Fluctuation in flow erodes the banks of the river, destroying shallow-water habitats needed for the hatching of eggs and nurturing of young. Moreover, the temporarily receding river exposes young fish; even those fortunate enough to find shelter in pools of water are vulnerable to predators.

The Role EIS admits that "the severity of the impact" on the Columbia River fisheries "cannot be determined."[15] Yet on the same page the assertion is made that "It is very likely that the net effect of hydropeaking on these populations will be unobservable."[16] No evidence is offered for this reassuring conclusion.

In 1977 a technical task force of the Pacific Northwest Utilities Conference Committee reported that the hydropower system would be unable to meet the forecasted loads in the region by the late 1980s. The PNUCC analysts estimated how much hydropower could be produced if river operations were subject to physical and today's regulatory constraints. Such an appraisal is related only indirectly to environmental quality considerations, since the appropriate regulatory arrangements for hydro-thermal peaking have not yet been designed. Nonetheless, they concluded that these and other preliminary studies have identified the following problems:

1. Excessive daily load swings
2. Large amounts of hydro energy required to replace thermal plants on forced and planned outages
3. Nighttime residual loads below minimum hydro generation requirements
4. Insufficient water and regulation capacity to carry sustained peaking
5. Intolerable fluctuations in pool tailwater levels.

No existing PNUCC entity appears suited to look into these problems.[17]

Multiple uses. The full development of the hydrothermal concept envisioned in the original Hydro-Thermal Program is incompatible with the multiple use traditional on the Columbia. In addition to the flood control problems mentioned above, irrigation, navigation, and recreation will all be affected adversely.

Thus far, irrigation water supply and energy for pumping have been met from the abundant flows of summer; the water itself has had few competing uses. In the future, however, the salability of hydro peaking power, even in the high-flow summer season, will be limited only by the capacity of the transmission network and the number of thermal plants to which the Columbia River hydro plants are connected. Irrigation will thus mean foregone power output as well as use of valuable pumping energy. One study estimates the value of foregone energy resulting from the proposed East High Division of the Columbia Basin Irrigation Project at $59 per acre per year, plus another $48 per acre for the value of the energy used for pumping. Since 6.5 million acres of land are under irrigation in the interior Columbia River system, the economic values implicit in these figures are comparable to the revenues now being derived from federal hydropower sales.[18]

As seasonal surplus flows and surplus energy disappear, the potential grows for sharp conflict over the cost of and subsidy enjoyed by irrigation, a conflict dividing the urban west side of the Pacific Northwest and the irrigated interior. There is little public understanding of the extent to which federal irrigation projects in the region are already subsidized by financial assistance from BPA power sales (roughly 85 percent of direct construction costs, varying with the project). The debate over federal power allocation and rates, combined with increased analysis of the direct and indirect costs involved, is likely to enhance public understanding. The cooperative federal development of power and irrigation, dating from the earliest activities of the Reclamation Service in the region, could turn out to be a casualty of the Hydro-Thermal Program.

The implications of peaking operation for navigation and river recreation are also unclear but probably significant. Navigation locks are installed at only the four lowermost mainstem Columbia River dams and the four mainstem Snake River dams. These dams will be the least affected by peaking operation at the Grand Coulee and Chief Joseph projects since the water level fluctuations will dampen out as they move downstream. Accordingly, interest in lower river navigation may lead to shifting the peaking burden to upstream plants such as Chief Joseph and Grand Coulee. Accordingly, lower-river navigation may not be directly affected by peaking, at least in the early years of hydro-thermal operation. It should be noted, however, that navigation is sensitive not only to water levels but to currents as well. Thus, the speed with which reservoir levels change can affect ship and barge maneuverability. In any event, since it is a congressionally authorized function of the downstream dams, navigation can eventually become a constraint on hydropower peaking.

Unlike navigation, recreation is only an incidental purpose of the projects involved and is thus without legal standing. About all that can be said is that recreational use will be difficult in the middle and upper reaches of the river, approaching the impossible as one nears the toe of Grand Coulee Dam. Fencing off the riverbank entirely is one of the alternatives under consideration for some of the upper reaches of the Columbia; it is a grim "alternative" for riparian wildlife.

The proposed conversion of hydro capacity to peaking operation means that it is now technically and economically feasible to expand further the hydroelectric generating capacity at existing dams. An additional 7,600 megawatts of hydro capacity are now authorized or under construction in the Federal Columbia River Power System.[19] Although the energy in the rivers of the Pacific Northwest is largely harnessed, the building of additional generating capacity permits more of that energy to be delivered in a shorter amount of time--the profile appropriate to peaking.[20] Given the capital investment already embodied in the federal dams, the addition of peaking capacity at existing installations is economically compelling. A third powerhouse structure has recently been built at Grand Coulee Dam, opening the way for augmentation of its capacity in the years to come. Moreover, with the recent rise in fuel prices and other costs of new thermal plants, it has now become economically feasible to build additional hydro projects in the water-rich Northwest. Hundreds of sites, many for pumped storage facilities, have been inventoried.[21] But there is considerable potential for environmental dispute in hydro project development, as the long controversy over the raising of Ross Dam on the Skagit River in northwestern Washington demonstrates.[22]

It is noteworthy, however, that for all the uncertainties of future operations--and the known environmental costs--hydropower is regarded by most environmentalists as less threatening than thermal energy supplies. For instance, we have found no evidence of conservationist opposition to the New Deal programs to develop the Columbia Basin.[23] Hydro is, moreover, applauded as a form of solar energy--as are the fossil remains of ancient plants.

Coal

Although coal remains the "ace in the hole" of the Carter Administration's tattered energy policy, there is still much to the sardonic adage, "there's only two things wrong with coal--you can't mine it and you can't burn it."[24] Coal-fired facilities south of Puget Sound in Centralia, Washington, emit sulfur oxides and other air pollutants which sour the rainfall in the Puget Sound region.[25] The problem of acid rain, which already causes widespread, economically significant damage in the Midwest and Northeast, is expected to accompany electricity derived from coal, despite significant advances in air pollution control technology. Air quality regulations sharply limit the development of coal-burning plants. In 1978 the U.S. Environmental Protection Agency denied a permit for the coal-powered Colstrip 3 and 4 facility in Montana, ruling that its air pollution would lead to impermissible degradation of Northern Cheyenne Indian reservation lands. The EPA action is being appealed, but the problems it reflects remain.[26]

Much of the coal needed to meet Pacific Northwest demand lies in the Rocky Mountain states, east of Northwest population centers. The opening of a thermal power station and associated strip mine in Rosebud County, Montana--haplessly named "Colstrip,"--has stirred bitter opposition among Northern Plains ranchers and Indians. The profitable but ugly stripmining apart, Montana landowners are furious that their Big Sky is being despoiled to provide a "boiler room for the Northwest states."[27] Indians, ranchers and environmentalists have been a principal source of demand that Montanans be granted a voice in regional decisions affecting their resources. In particular, Montana politicians have insisted on full representation in Pacific Northwest power planning, even though only the western part of the state lies within the West Group area.[28]

Even if the massive scarring of land, the boom-town disruptions of forced-draft industrialization,[29] and the pervasive burdens of air pollution were all overcome, coal would still pose a sizable challenge in the Rockies: the water supply is variable and there is little to spare. Thermal plants must have water for cooling, some of which is lost by evaporation unless expensive dry cooling towers are used. Stripmined land requires steady, heavy applications of water if it is to be reclaimed to the standards now required by federal law. Existing users have already claimed more than the available supply, and the antiquated, intricate law of water rights poses complex and litigious barriers to redistribution or even sale of water claims. A recent analysis by Harte and El Gasseir concludes that "production of steam-generated electricity would cause conflicts in the use of fresh water unless dry cooling were extensively used."[30]

Nuclear

In comparison to the known deleterious effects of coal, the environmental debits of nuclear energy were, until the Three Mile Island accident, fears about the future. But the fears are awesome and apocalyptic, and they cannot be dispelled by experts' reassurance, if recent history is a reliable guide. Even President Carter, once a nuclear engineer of sorts, calls atomic power a last resort--if not the ultimate weapon in his moral equivalent of war. Five issues cloud the future of nuclear energy: routine releases of radioactivity; the possibility of catastrophic accident; concern that commercial nuclear power development may lead to the spread of nuclear weaponry to other nations and terrorist groups; radioactive waste disposal; and skepticism about the reliability and economics of nuclear power. We have listed these in order of increasing political gravity; in combination, they raise doubts about the survival of the U.S. nuclear power industry that can no longer be dismissed.

Hazard. Catastrophic accidents and routine releases of radioactivity most readily capture the popular imagination. Though controversy has been fanned by the Harrisburg accident, both hazards have been mitigated considerably by sizable investments and conservative safety margins.[31] "No one died at Three Mile Island," as the defenders of nuclear energy proclaim, and the official estimates of radiation exposure indicate that future damage to the exposed population is for all practical purposes negligible.

What did suffer a life-endangering blow in March 1979 was the credibility of the nuclear industry, especially its smug reliance on technological safety systems. According to the investigations underway, these systems seem to be neither well-designed nor invulnerable to bureaucratic bungling or plain human carelessness. Moreover, the industry must contend with the national policy of slowing the proliferation of nuclear weapons. Both U.S. export restrictions and the halting of plans to reprocess spent nuclear fuel[32] deepens the pall hanging over the industry.

Management of nuclear wastes is a Northwest issue, since most of the waste from the military use of nuclear energy--much of which stems from nuclear propulsion of submarines and ships as well as weaponry--is stored at the Idaho National Energy Laboratory and at the Hanford Reservation in Washington. This "temporary" storage continues to be, despite some operational problems, the principal stopgap means of nuclear-waste management. Although officials at the U.S. Department of Energy are pursuing research relating to nuclear waste, they concede that another twenty to thirty years may be necessary before full-scale implementation of an ultimate disposal mechanism can be undertaken.[33]

Like proliferation, waste management does not affect directly the development of nuclear energy in the region. But the fact that there is no federal policy for disposal of radioactive wastes has become a serious liability. One member of the Council on Environmental Quality has suggested that no new nuclear power stations be permitted to operate until the federal waste management situation is cleared up.[34] A federal court has implicitly agreed with this position with respect to the Vermont Yankee power plant in New England.[35] And in 1976 California passed a law requiring new power stations to provide satisfactory evidence that waste management problems would be resolved;[36] other states are following suit. No new nuclear plants have been approved in California since 1976.

Controversy. These four environmental criticisms of nuclear energy have led to a spirited public debate, one which has contributed to changes in nuclear policy at all levels of government. As a recent defense of nuclear power in *Time* magazine illustrated, moreover, these concerns have set the agenda of public discussion.[37] As a consequence, nuclear power is decidedly on the defensive, and the industry finds itself in the unenviable position of "proving a negative"--that nuclear energy is safe, even though the scientific evidence still indicates its hazard levels are extraordinarily low when nuclear is compared to other technological undertakings.

Significantly, every attempt to put nuclear power to a popular vote has led to substantial--if far from unanimous--support for nuclear energy; the siting of nuclear facilities, especially for radioactive wastes, has been accorded a cooler reception in referenda. Antinuclear initiatives on the state ballots of Oregon and Washington have been defeated, though an initiative strengthening review of nuclear plant siting in Montana passed.[38] The California vote was preceded, however, by the enactment of stringent legislation which has, as noted above, effectively restricted the development of nuclear power in the state.

Northwest nuclear projects have been buffeted by considerable controversy, except for those located in the Tri-Cities area of south-central Washington. There, the prosperity brought by the Hanford Reservation has fostered enthusiastic advocacy of nuclear power, nuclear plants, and even disposal of nuclear wastes in the underlying basalt rock. Elsewhere, controversy reigns.

The Pebble Springs nuclear project in Oregon was challenged in litigation that reached the Oregon Supreme Court; in 1977 the Oregon Department of Energy entered the lists, suggesting that the need for the Pebble Springs plant had not been adequately argued by its principal sponsor, Portland General Electric. In 1978 the Oregon DOE expanded its challenge, questioning the economic and operational benefits of nuclear as compared with coal; the regional strategy of emphasizing baseload plant development; and the planning framework for regional energy investment planning.[39]

Since 1974 the beautiful Skagit River Valley near Sedro Woolley, Washington, a rural community in the Cascade foothills, has been the center of a persistent trench war. Puget Sound Power and Light proposes to build two nuclear plants here. The utility has sought to win the support of the plants' neighbors by agreeing to construct fish hatching facilities and recreation areas, and by committing itself to help pay for the social services and other costs imposed on the community by plant construction. "One Nuke Gets a Warm Welcome," one article on the proposal concluded optimistically.[40] Although Puget Power managed to get state approval of the Skagit site in 1976, the proceedings before the U.S. Nuclear Regulatory Commission to obtain a limited work authorization permit have dragged on, with no clear sign of resolution in mid-1979.

To a large degree, the quagmire encountered by Puget Power has been engineered by one man, Roger Leed, an iconoclastic Seattle attorney who represents Skagitonians Concerned About Nuclear Power.[41] Leed has pursued every loose thread in the utility's argument, raising questions about plant safety, environmental impact, the need for power, and--most effectively--possible seismic hazards at the plant's location in the Skagit Valley. A fault has now been found close to the plant site, though whether it has shown signs of activity recently enough to be considered hazardous is yet unclear. The battle is one where every delay is a defeat for the plant's proponents, especially in light of the rapid cost increases now plaguing the industry.

Until 1978 the Washington Public Power Supply System had remained remarkably free of controversy in pursuing the construction of five nuclear plants. Three are at the Hanford Reservation, and two others are located southwest of the lower tip of Puget Sound, in an agricultural valley near the village of Satsop, Washington. Although there were local protests entered at the state siting council hearings for the Satsop plants, the requisite permits were granted by the state and federal agencies.

Satsop has recently become a target of the Crabshell Alliance, a loose-knit group of nuclear protestors who are protesting both the nuclear plants and the Trident submarine base under construction at Bangor, Washington. The base will house the nation's newest fleet of ballistic missile-firing boats beginning in the early 1980s. The Crabshell protest thus aims at linking in the public mind military and civilian uses of nuclear energy. In a series of demonstrations at the Trident base and the nuclear plant site, several hundred Crabshell Alliance members were arrested in the spring of 1978. Their civil disobedience protests were so well-planned that the Navy substituted box lunches for billy clubs: each of the arrested demonstrators found a governmentally approved nutritious meal on the seat of the bus waiting to take the "criminal" trespassers to the local detention center. It is hard to say how much the Satsop plants themselves will be affected by this media-oriented style of political action, aimed as it is at a much wider audience and a much grander agenda.

Economics. It is the fifth of the difficulties plaguing nuclear energy, cost and reliability, that in the short run most threatens atomic power in the region. The problems besetting the Trojan plant and the five plants under construction by the Washington Public Power Supply System illustrate the issues. Trojan's operating history since its completion in 1975 has been quite uneven. The worst episode began in March 1978, when the reactor was shut down for refueling. An inspection of the control building revealed that it did not have the protections from earthquake movement required by federal regulations. The U.S. Nuclear Regulatory Commission (NRC) halted resumption of power operations pending resolution of this problem, and Portland General Electric (PGE) filed suit against Bechtel Corporation of San Francisco, which had designed and built the control building.

By the end of 1978 PGE had lost its suit, purchased over $20 million of power from other sources to replace Trojan's output, and corrective measures were underway. The Eugene Water and Electric Board (EWEB) owns 30 percent of Trojan, financing its participation through the net-billing arrangements of the Hydro-Thermal Program.

As a result, Bonneville--not EWEB--became liable for meeting the capital obligations of EWEB's portion of the plant when it was shut down. The region's consumers--mainly the preference customers--have thus shouldered the $2.6 million monthly share of Trojan's closure; PGE is liable for roughly twice as much, although its bill will not fall due on the same schedule because of differences in financing arrangements.

The 1978 Trojan episode is an isolated worst case. It illustrates, nonetheless, the complexity of nuclear plant operation and its tight regulatory framework. But in general the reliability of nuclear power stations, especially newer plants larger than 600 megawatts, has been disappointing.[42] With lower reliability the availability of plants to produce electricity has remained low. The cost of power has climbed as a result, for the two reasons suggested by PGE's experience. First, replacement power to serve the utility's load is usually more costly than the nuclear plant's electricity. Second, the capital costs of the nuclear power plant need to be paid anyway, since these are financed by external borrowing that comes due even if the plant is temporarily unproductive.

It should be no surprise that the disappointing performance of completed nuclear plants is matched by delays and problems in the construction of new plants. At times, as with Puget Power's frustrations with the Skagit project, the slowdowns result from environmental opposition before regulatory agencies.[43] But the rapidly rising cost of nuclear energy is basically pushed by wage increases, fuel and material supply uncertainties, shifting government regulations, and management overmatched by the sheer complexity of nuclear enterprise.

The best-known case in the Northwest is that of WPPSS.[44] As the leading regional builder of nuclear plants, the Supply System is a primary implementation agent of the Hydro-Thermal Program's thermal development. When its five nuclear plants are completed in the mid-1980s, WPPSS will have a generating capacity close to half that of the Federal Columbia River Power System.

The enormous scale of the WPPSS undertaking has proved hard to sustain; costs and delays have mounted relentlessly. Although the figures change monthly, it is clear that the WPPSS plants will be well over twice as expensive as originally planned (Table 11). In all, the five plants have experienced delays of over fourteen years. And the cost *increase* in the five WPPSS projects exceeds the *total* capital investment allocated to electric power development in the federal system.[45] Although the Supply System's costs are discouragingly high

Table 11. *WPPSS Nuclear Projects Capital Cost Comparison ($ in millions).*

Project	Location	Announcement	Estimate at first permanent financing	Estimate in July 1979
1	Hanford	$ 632.9 (1972)	$1,204.0 (1975)	$ 2,341.4
2	Hanford*	384.9 (1970)	504.0 (1973)	1,821.7
3	Satsop	755.7 (1973)	1,402.0 (1975)	2,256.2
4	Hanford	1,009.0 (1974)	1,610.3 (1977)	2,580.0
5	Satsop	1,210.0 (1974)	1,963.0 (1977)	2,753.0
TOTAL		$3,992.5	$6,556.3	$11,752.3

*WNP2 was originally sited at Roosevelt Beach, in Grays Harbor County.

Source: Washington Public Power Supply System.

in the hydro-rich Northwest, its experience is not unique, as Figure 15 shows. Nuclear energy is costly anywhere; its price is rising everywhere.[46]

Despite the difficulties of nuclear and coal technologies, both retain strong utility support. Both use familiar contractual and engineering methods: wind does not. The technocratic decision making discussed in Chapter III biases utilities toward large, centralized generation technologies.[47] Some of the alternative means of meeting electrical demands imply a shift away from electricity, as with conservation or solar space and water heating. Others, like biomass, wind or low-head hydro, utilize generation techniques not engineered for use at the industrial scale traditional in regional utility planning; some, like cogeneration, are inappropriate for large-scale, central-station concepts. Yet the economic meaning of higher energy costs in petroleum, thermal electric, and hydro is that sources once considered too expensive, including conservation, now need careful re-examination. Much is now economically feasible, including some technologies whose organizational implications may be inimical to existing utilities.

Regional Impacts

In addition to direct environmental impacts, there are a number of indirect impacts and related develop-

Figure 15. *Completion dates and Approximate Costs of the Supply System's Plants Compared with those of TVA.*

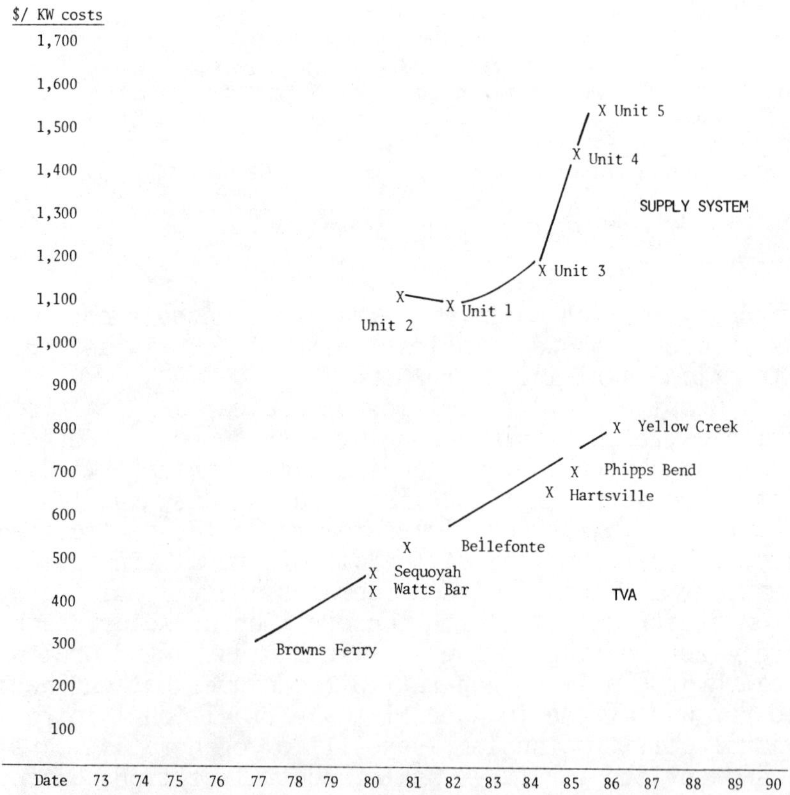

Source: Dan Seligman, *The Washington Public Power Supply System*, unpublished report, July 1978.

ments that affect the future of Northwest power development. Although these region-wide factors do not have clear or simple connections to utility plans, public perceptions regarding their importance are likely to influence the political response to regional electric power. Three impacts are of special prominence: the possible development of a regional nuclear energy center; the possibility of major exports of power; and the general trend toward urbanization, particularly in the western part of the Pacific Northwest.

Energy Centers. The presence in the Pacific Northwest of a major nuclear waste storage facility, the relative abundance of cooling water for the operating of nuclear plants, low population density--thus easier siting --and the flexible hydro resource all make it sensible to think of the region as the location for one or more nuclear energy centers serving the western United States. Nuclear energy centers, currently under consideration by the U.S. Nuclear Regulatory Commission, are facilities with generating capabilities ranging from 10,000 to 40,000 megawatts--power parks in which up to forty nuclear plants could be located in a relatively compact area.[48] The Hanford Reservation already contains roughly a dozen nuclear reactors, all inactive remnants of military nuclear material production except the four WPPSS facilities. Hanford seems destined for additional commercial nuclear plants. Moreover, Pacific Power and Light Corporation, one of the region's investor-owned utilities, has already initiated discussions with Washington's Energy Facility Site Evaluation Council about the feasibility of building a ten-plant coal and nuclear energy center near Hanford.

Exports. The nuclear energy center concept raises broader concerns about the Northwest's relationship with other regions in the country, especially the Southwest and the large electricity markets of California and Arizona. Certainly it seems feasible from a technological--if not environmental--standpoint, to combine the Northwest's abundant water resources and the history of atomic power development in central Washington to build a major generating complex. Moreover, if the land use and environmental problems of strip mining in the arid West are not satisfactorily resolved, it is possible that additional coal-fired power may be denied to the large, and still growing Southwest markets. In that context, and given the economic and political clout of the Southwest, a national policy calling for the harnessing of Pacific Northwest resources to supply electric power south seems, if controversial, not impossible.[49]

Although built in part to transport Canada's share of Columbia River Treaty power to the Southwest, the Pacific Intertie has been used principally to exchange electricity between the Northwest and Southwest.[50] The Intertie could in future be the basis for larger interregional transfers of electric power. An export economy, however, would necessitate enlarging the transmission capability. Such a project would require cooperation among utilities and government, cooperation which the regional power debate may have made problematic.

Urbanization and Growth. The scale of the Hydro-Thermal Program is consistent with, if not actually abetting, increased urbanization and industrial development throughout the Pacific Northwest, especially in its western third. Growth, which still enjoyed boosterish support only five years ago, is now commonly treated as an unwanted inevitability, or worse. "Have you been rude to a tourist today?" demands a Seattle bumper sticker in posh Madison Park, a neighborhood where much of the city's business elite resides.

The corridor formed by the Willamette Valley and the Puget Sound lowlands running from Eugene, Oregon, to Vancouver, British Columbia, presently houses most of the population in the region. This strip, 50 miles wide and 350 miles long, is far from being a megalopolis of the sort found in the Northeast or the Great Lakes region. Yet its trend has been toward increased urbanization and industrialization. With regional power development and the prospect of a major influx of Alaska North Slope petroleum into Puget Sound, one observer has noted acerbically that this corridor might become the "New Jersey of the West Coast."[51]

Growth pressures, however, are likely to result in differing responses among the states of the region. Oregon has already enacted strong land-use regulation --partly because its growth has been the most rapid in the region.[52] Idaho, whose growth rates are approaching those of Oregon, is still pursuing expansion. Washington has vacillated. Such differences at the state level are likely to influence future patterns of development. Correspondingly, the evolution of growth policy and growth itself are likely to have different impacts on political perceptions of the regional power controversy.

Under existing policy, the impact of higher electricity costs will fall unevenly on Northwest consumers. As a result, language proposed in congressional legislation has reflected three divergent positions. Economists, environmentalists, and those from slow-growing areas like Seattle advocate economic incentives in power pricing that reward efficient use and protect those that grow least. Those from rapidly growing areas--mainly suburbs and some agricultural areas where irrigation is expanding--demand lower prices for "rural and domestic customers," who were arguably accorded preference in the Bonneville Project Act. And public agencies, who enjoy preference now, argue their historical claim, backed by the political clout of the well-organized national public power lobby.

Growth is a confounding political issue. A number of public agencies serve rapidly growing areas, although growth patterns do not clearly favor public or private service areas. The eastern parts of the region, traditionally conservative and favorable to growth, fear their low-cost irrigation power may be priced out of reach by urban sprawl in the liberal western areas --a fear that ignores the rapid expansion of irrigation pumping in eastern agricultural areas. And those living and working near major power facilities like Colstrip, Skagit, or the Hanford Reservation have intense feelings against or in favor of plants whose economic reach is regional or even national. All these factors cut across the usual political alignments, increasing the confusion attending the complexities of regional power development.

The most recent public opinion survey devoted to examining these issues was done in 1975 for the Pacific Northwest Utilities Conference Committee.[53] It showed that the utilities' forecasts of the need for power were declining in credibility, with half the populace refusing to believe in a future shortage of electricity. In addition, half the sample favored conservation as a way to meet power needs. Support for nuclear power, still the first choice of 38 percent of the sample, was declining, while solar, already the favorite of one-fifth of the people, was on the rise. "Demographically, young adults...invariably adopted the environmentalist point of view," the pollsters told the utility managers.[54]

More broadly, a 1978 national survey conducted for Resources for the Future showed that "support for environmental protection is strong and unwavering...with no sign of a backlash.[55] Indeed, air and water pollution each ranked with energy problems and unemployment in importance, with nearly one-third of the sample agreeing that these were "very serious" problems in the nation. Sixty percent of the population is either sympathetic to or an active participant in the environmental movement, and 29 percent are inclined to oppose nuclear energy. The opinion base of support for the broadened agenda of debate is, by the standards of public attitude studies, strong and substantial.

RESPONSES

The agenda of public discussion about regional power development grew slowly at first. It began in 1973 as an attempt to strengthen the Hydro-Thermal Power Program by revising the financing arrangements for thermal power stations. Three years later, as the 1976 election brought an environmentalist to the White House, HTPP lay in shambles. Long-held assumptions about the inevitability of growth, the tolerability of environmental impacts, and the actions of the utilities were being subjected to close, skeptical analysis; and the region's leading public utility, Seattle City Light, had spurned participation in two WPPSS plants. The can of worms was open.

Phase 2

As described in Chapter III, the net-billing arrangement had a basic limitation in the amount of financing that could be supplied. At any time, the annual revenues paid to Bonneville by the public utilities for firm power set a limit to the amount BPA can offset by net billing. But the cost of thermal power plants rose so rapidly that it took larger and larger quantities of federal hydropower to offset the cost of each unit of thermally generated electricity. Seeing that the net billing capability would be swiftly consumed by high-cost thermal generation, Bonneville announced in December 1973 an end to net billing for new plants and unveiled plans for Phase 2 of the Hydro-Thermal Power Program.

Phase 2 plans were made in the face of rising costs and delays in thermal projects. The delays decreased the amounts of power available in the near future, leading to fears of power shortages by the early 1980s; although droughts and recession have slowed demand, delays have grown too, and fear of shortages remains a dominant factor in utility planning today. To minimize delay, several of the power plants proposed in Phase 2 use engineering designs developed during the original Hydro-Thermal Program, now renamed Phase 1.[56]

Financially, Phase 2 was built around three central provisions.[57] The first two replaced the net-billing arrangement with an alternative means for financing new thermal power plants. Bonneville would return to being a wholesale marketing agent without acquiring thermal project output as it did in Phase 1. In what amounted to a resumption of the partnership arrangements of the 1950s, new thermal plants would be financed directly by sponsors such as WPPSS without BPA backing. Sponsors would bear the risk of unsatisfactory performance. The projects' sponsors would execute with the utilities wishing to purchase power a set of participation agreements, fiscal instruments which would have the effect of aggregating the advantageous borrowing capability of the public agencies.[58] Supplementing these arrangements with the publicly owned utilities would be a new set of contracts with industrial customers. None of these three elements of Phase 2 has been implemented. Yet Phase 2 is crucial to an understanding of the interdependence among utilities, BPA, and the major industrial customers, an interdependence that has continued to guide them.

If BPA's financial role was to diminish under Phase 2, its managerial oversight would have been enhanced. Bonneville was to have entered into "trust agency" agreements with its preference utility customers. As the trust agent of the utilities, BPA would represent the public systems in dealing with thermal project sponsors, including sitting on boards of directors and other institutionalized forms of oversight. Already, BPA's oversight role, based on its role in financing net-billed facilities, has become controversial.[59] Bonneville inquiries into the rapidly escalating costs of WPPSS Nuclear Projects 1, 2, and 3, the three net-billed facilities, has been regarded as meddling by WPPSS officials. A trust agency role would have given BPA similar or enlarged authority with respect to Projects 4 and 5, which now lie outside Bonneville's purview.

Augmented BPA responsibilities to represent preference customers would be balanced, however, by the utilities' formal powers to restrain Bonneville's interpretation of their interests. A trust agency council, resembling utility advisory boards later incorporated in congressional legislation, could veto BPA decisions to market the power from new generating facilities.

Another feature of Phase 2 was the redefinition of the regional role of the direct-service industrial customers. The DSIs were offered the new IF-1 rate schedule, the "industrial firm" power described in Chapter II. Less a redefinition than a formalization of emerging relations between Bonneville and the DSIs, IF-1 explicitly assigned the burden for delays in the construction of net-billed plants to the industrial customers.[60] In return, the DSIs were to receive new twenty-year contracts, guaranteeing service into the 1990s.

Phase 2 began smoothly enough. In 1974 Congress passed the Federal Columbia Transmission System Act (P.L. 93-454), a piece of regional "housekeeping" legislation that contained an important new authority for BPA: self-financing. The act gives Bonneville control over the revenues obtained by its sales, together with limited borrowing capability through bond sales. Freed of the necessity to secure appropriations for short-run expenditures, Bonneville can now use these revenues to augment the regional transmission grid in support of regional power development. This did not, however, enable Bonneville to finance major expansion. Still, preliminary planning for the WPPSS nuclear plants--which were not covered by net billing--proceeded smoothly toward financing. WPPSS Nuclear Project 4 (WNP 4) was sited at Hanford next to WNP 1, its "twin" plant. This promised significant savings since WNP 4 would use a common design and common facilities whenever possible. WNP 5 was located at Satsop and matched to WNP 3 there.

Phase 2 ran into trouble in 1975. Fittingly, the National Environmental Policy Act (NEPA), with its famous requirement for environmental impact statements, was the instrument. But in an irony that would have amused the New Dealers who fostered economic development in the Northwest, the plaintiffs were not Hodel's reviled "prophets of shortage," but a determined band of small-town businessmen anxious to bring a lucrative --if polluting--industrial plant to their town.[61]

In 1966, plans to launch a new direct-service industrial customer began to take shape with the signing of a power sales contract between BPA and the Northwest Aluminum Company. Over the next nine years that contract changed owners three times (once only nominally) and the location of the proposed plant changed too. Its penultimate proposed site was Warrenton, Oregon, at the mouth of the Columbia River in Clatsop County.

There the plant encountered environmental opposition from the local Clatsop Environmental Council, the Oregon Environmental Council, and the Northwest Environmental Defense Center. These organizations filed suit in 1974, claiming violations of NEPA and the Clean Air Act. Even before the suit was filed, the Oregon Environmental Quality Commission had suggested that the Warrenton site would not receive air and water quality permits.

Bowing to these pressures, the plant's sponsor, Alumax, announced that it would change the location to Umatilla, in eastern Oregon. As a condition of shifting the point of delivery, Bonneville asked Alumax to sign one of the Phase 2 IF-1 contracts. Alumax was the first DSI asked to do so. It was to be the last for a good long while.

For at this point, in 1975, the Port of Astoria, a public authority in Clatsop County, and a group calling itself "Concerned Citizens of Clatsop County" filed suit, naming BPA administrator Don Hodel as defendant.

Port of Astoria v. *Hodel* turned on whether the Alumax contract for IF-1 power constituted a major federal action requiring a full-scale environmental impact statement.[62]

The U.S. District Court ruled that it did, and further, that the EIS must include an examination of Phase 2. This was the beginning of the five-volume BPA draft, "The Role of the Bonneville Power Administration in the Pacific Northwest Power Supply System, Including Its Participation in the Hydro-Thermal Power Program." The Role EIS was the end of Phase 2, since none of the contractual superstructure of the Hydro-Thermal Program could be implemented until the final impact statement was issued, an interregnum which so far has lasted five years.

Even without the doughty burghers of Clatsop County, Phase 2 may have been doomed. For in the docket a few cases behind *Port of Astoria* was *Natural Resources Defense Council* v. *Hodel*, an environmentalist challenge to Phase 2 seeking the relief granted in *Port of Astoria*.[63] The NRDC suit was suspended pending completion of the Role EIS.[64]

Whether the enemy were environmentalist-prophets or businessmen in pursuit of the profits of growth, the Hydro-Thermal Power Program had become a victim of the increasingly visible social significance of electric power development. Like other undertakings of industrial technostructure, the ability to plan was central to HTPP. And the interruption created by *Port of Astoria* destroyed that ability.

Energy 1990

As part of the implementation of Phase 2 of the Hydro-Thermal program, the Washington Public Power Supply System began in the spring of 1975 to arrange financing from Northwest utilities for two nuclear plants, WPPSS Nuclear Projects 4 and 5. WNP 4 is located at Hanford, WNP 5 at Satsop; neither plant was financed through net billing. As the largest municipally owned utility in the region Seattle City Light (SCL) was asked to participate in both plants, to meet anticipated growth in its load. The SCL staff proposed that the city purchase a 10 percent share of each facility.

Nearly a year and a half later, in July 1976, the Seattle City Council, on a six to three vote, voted against participation, and moved instead toward an ambitious program of electric energy conservation which will, over the next decade, attempt to slow growth in demand sufficiently to obviate the need for participation in the WPPSS projects. The interaction of utility, business, governmental, environmental, and consumer interests in the Energy 1990 decision (as it came to be called) exemplifies the broadening of the agenda of public debate, a broadening which would later thwart attempts at congressional legislation. Our account of Energy 1990 is drawn from a study by Greg Hill, "The Politics of Energy 1990."[65]

Participation in WNP 4 and 5 was estimated to cost more than $250 million. In the spring of 1975 the City Light staff sought approval from the Seattle City Council of an option agreement--a preliminary commitment which would enable Seattle to participate in the power plants, but would not commit the city to a final decision for several months. The council approved entering the option agreement, but requested that City Light undertake with the council a study of the implications of so large an investment. The option agreement proposal prompted a suit by the Washington Environmental Council, an umbrella group of environmental organizations in the state. The suit sought to halt Seattle's participation pending the filing of an environmental impact statement pursuant to the Washington State Environmental Policy Act. The Environmental Council suit was withdrawn as part of a negotiated agreement with the city that the scope of the study requested by the city council would satisfy the environmental disclosure requirements of the law.

The resulting study, entitled *Energy 1990*, took nearly a year to write, and subsequently became the focus of heated political debate.[66] Over 100 newspaper articles, as well as a number of television and radio programs, were written about Energy 1990 in the fifteen months between Seattle's signing of the option agreement and its decision to withdraw from participation.

Energy 1990, when it was issued in February 1976, defined three issues which have continued to dominate the regional power controversy. First, the study made explicit to political leaders the drastically higher costs of thermal power; correspondingly, the economic benefits to be reaped through conservation--"non-generation alternatives" in the parlance of the study--also became apparent. Second, shared public values and the responsibilities of government to implement those values became the focus of sharp debate; was City Light's responsibility to meet demand, or was it instead the responsibility of Seattle's elected government to impose policies which might maintain as long as possible the low cost of electricity?

Third, a critical issue of methodology affected both the salience of costs and the perceptions of public values. This was the testing of electrical load forecasts in the fires of political debate. One of the city government's analysts thought that "the forecast would settle it."[67] But as estimates of future consumption became entangled with value judgments of what kind of future was desirable, and the nine-member city council was drawn into the role of key decision maker, one of the council members observed wryly that "the electrical growth rate will be what five votes says it is."[68]

The credibility of the forecast prepared by an independent consultant for *Energy 1990* turned on its persuasive use of econometrics, a combination of economic and statistical analysis, instead of reliance upon the extrapolation of historical trends. The critical advantage of econometric analysis is that it can take into account--if imperfectly--the possibility that rising prices will affect the amounts consumed.

In the event, the *Energy 1990* forecast has proved to be pleasantly mistaken. By the 1978-79 "water year," the estimated energy demand projected in the forecast was 1,031 average megawatts; the actual consumption, boosted by a robust economy and an extraordinarily cold winter, was only 956 megawatts.[69] Whether this relatively low figure is attributable to conservation is less certain. An improved building code, mandating energy-saving designs and practices, has been delayed by wrangling between the city council and the mayor's staff. These tensions came to a head in the bitter battle over the confirmation of conservation-minded engineer Robert Murray as superintendent of City

Light. Although Murray survived a bruising debate in the spring of 1979, his ability to work smoothly with the council remains to be tested.

The *Energy 1990* study was prepared by a group of independent consultants for the City of Seattle and Seattle City Light. The selection of the consultant team and review of its performance were conducted by citizen panels. A month after the study was released, a public hearing before the council, sitting as a committee of the whole, was held on the report.

The comments of Seattle residents made it dramatically clear that concern about rising costs of electricity was far more persuasive than the dire predictions of possible electric power shortages by utility experts. This emphasis was shared by a two-thirds majority of the Citizen Overview Committee (COC) which monitored the study. Eighteen of the twenty-seven members of COC recommended against participation. "The COC majority ...was not only willing to accept the risk of future shortages, it was willing to rely on that risk as a necessary motivation to conserve."[70] The recommendation was given added public credibility by the fact that one of the leaders of the COC majority was Dr. Gene L. Woodruff, professor of nuclear engineering at the University of Washington. Woodruff was a strong supporter of nuclear power, but he was nonetheless persuaded that the savings from feasible conservation programs were economically superior to the high costs of WNP 4 and 5.

The COC recommendations outlined a demand management policy and, significantly, supplemented the work of the independent consultants with additional information about feasible conservation measures. As Hill observes, the information developed by the COC majority about conservation altered the definition of the social purpose underlying parsimonious energy use: "The appeal to a new 'energy ethics,'...was withdrawn and in its place was substituted a collection of cost-effective conservation measures."[71] Conservation could claim to be a conservative program, not just the moralistic regulatory strategy of liberal environmentalists.[72]

The independent study and its citizen overseers did not, however, convince the utility regulars, who were mindful of Seattle's leading role in the regionwide thermal construction program. City Light superintendent Gordon Vickery recommended to the city council that Seattle participate in WNP 4 and 5 anyway. As a conciliatory move, he advocated participation of only 5 percent in each plant, rather than the 10 percent reserved under the option agreement. In addition, Vickery recommended development of other resources, together with conservation. He had spurned the outsiders' analysis. Mayor Wes Uhlman, facing an uphill primary battle for the Democratic gubernatorial nomination, ducked the issue by agreeing with City Light's recommendation, but conditioning it upon standby contracts to market any surplus from Seattle's share of WNP 4 and 5 at the full costs of production. (Uhlman's attempt to steer a moderate path between environmentalist Marvin Durning and nuclear advocate Dixy Lee Ray did not, in the event, bring him enough votes to overtake populist-conservative Ray.)

The city council was thus faced with an executive branch recommendation apparently insulated from economic liability by the standby sale agreements. Yet, dubious that power from WNP 4 and 5 could compete reliably against other sources, including secondary hydropower when available, the council swung back toward the position advanced by the non-governmental analysts. The council voted by the surprisingly large margin of six to three against participation and for investments in conservation. Hill concludes that the pivotal factor in the decision was that participation in the nuclear plants would have increased costs to residential users more than conservation programs, even mandatory ones.[73]

The "1990 paradigm"

Seattle's fifteen-month decision process paralleled the stalling of Phase 2 of HTPP. And, more clearly than the difficulties encountered by the Hydro-Thermal Program, Energy 1990 made dramatically clear the way in which the politics of electric power had changed in the Pacific Northwest.

First, the *dramatis personae* had changed. City Light had introduced the power plant investment, with the backing of the business establishment, as a necessary expansion of SCL's generating resources. But the development of environmental controversy brought the decision into the public arena. There, mounting unease by consumers, principally residential users of electricity, grew rapidly, eventually looming larger than the claims of the well-organized environmental movement. As in the regional debate, environmentalists served as a catalyst, slowing the momentum of thermal power development, permitting doubts about the economics of thermal energy to gather force.

With the rising controversy, government could no longer ratify utility recommendations as usual, but had instead to adjudicate among competing claims. The executive branch, including City Light, tried to formulate a workable compromise: participation at a lower level in the nuclear projects, coupled with assurances that surplus power could be sold at or above its cost to the city. The legislative branch--whose authority was determinative--balanced the interests differently, spurning the assurances of the utility and the political judgment of the mayor's office. Their most vocal constituency, residential ratepayers who had expressed concern about rising power costs, needed conservation, not thermal plant investments.

Not only the actors, but their tactics were different.

> These "Energy 1990" decisions were remarkable in a number of aspects: together they amounted to a new energy policy which emphasized the management of demand rather than the continued increase of supply...; the decisions were arrived at following a year-long process in which an independent energy forecast and various policy alternatives were developed and debated within a setting of extensive public involvement; and, they were arrived at despite the persistent opposition of some of the most powerful interests in the city.[74]

The dynamics of Energy 1990, moreover, capture in microcosm the evolution of regional power issues. The proposal to participate in thermal generation arose amid forecasts by utility experts of distant, but economically threatening power shortages; a stalling lawsuit was filed, then withdrawn in favor of a study conducted by outsiders, with citizen oversight, and within the framework of an environmental impact statement report; policy was then made authoritatively by a legislative body, in the face of considerable politicking by the executive branch; and, as is often characteristic of legislative policy, subsequent implementation of policy by the executive branch has been uneven, even weak.

The reliance on legislation and expertise--manifested in the econometric techniques used in the consultant forecast--recalls an earlier period of American political life, the Progressive Era.[75] The Progressive movement of the early twentieth century was as complex in its social origins as any contemporary political shift. It seems, nonetheless, to have been part of a broader realignment in response to the urbanization and industrialization of American cities. In particular, a new class of middle-class professionals began to assert its influence in the Progressive reform movement. In parallel, one might identify the pivotal role of affluent urban and suburban professionals in much of the "public interest" movement of the 1970s, including environmental, consumer, and social welfare advocacy.[76]

At the same time, the resurgence of legislative decision making may be interpreted as a recognition of two factors. First, that public decisions need to reflect the divided power that is concerted in legislative action. Second, that large-scale technological decisions, such as commitment to thermal power development, have long-run but poorly seen social effects, and thus amount to legislative social choices despite their nominally technical character.[77]

These parallels with Progressivism reflect the fact that the social conditions of the Energy 1990 decision resembled those of the turn-of-the-century urbanization. The possibility of mobilizing middle-class citizens, combined with the need to build legislative coalitions, made the circumstances ripe for political activity by interest groups. Most of the Citizens Oversight Committee members were affiliated with organizations visible in the city, several of them citizens' groups. The COC majority "formed a classic progressive coalition--environmentalists, academics, community councils, the League of Women Voters, People Power,

Metrocenter, and the Municipal League--the sort of middle-class groups that were coming to exercise more and more power in this middle-class city."[78] The political tenor of this coalition decisively affected politicians' perceptions: Energy 1990 "shaped up as a battle between 'the interests' and the City's well-coordinated voluntary associations.[79]

If this control of political perceptions proved to be decisive, the decision style evolved during Energy 1990 was rather more traditional. Billed as a struggle between utility "experts" and "citizens," Energy 1990 had plenty of expertise speaking on both sides--notably in the COC's work in developing information about cost-effective conservation. The abundance of expertise did not, however, eliminate the need for pragmatic compromise. This was so even when expert knowledge seemed clearly to be at stake, a circumstance in which a more collegial, scientific method might have been expected to prevail, especially among citizens cautious about politics-as-usual.[80] Certainly the COC played considerably more than an advisory role, especially after the econometric study was completed.

The way in which environmental opposition provided the catalytic delay which permitted concern about costs to gather political momentum has been misunderstood and repeatedly misrepresented by members of the traditional utility-industry-government establishment. By singling out environmentalists, these spokesmen of the traditional order have spotlighted the power of judicial challenges to halt temporarily large technological systems in development by making orderly planning impossible. Such delays would, however, be merely temporary, given the significant but hardly overwhelming political strength of environmentalists, once a political showdown is pressed. What holds that showdown at bay is the belief that it would bring to the fore consumer concerns about rising power prices, a voters' displeasure that governmental decision makers cannot afford to unleash injudiciously. The power of environmental opposition thus does not reside in support for environmental quality itself, but in the economic implications of the technological transformation underway in the Pacific Northwest.

CHAPTER V
The Dance of Legislation

> Every statute may be said to have a long lineage of statutes behind it; and whether that lineage be honorable or of ill repute is as much a question as to each individual statute as it can be with regard to the ancestry of each individual legislator. ... Once begin the dance of legislation, and you must struggle through its mazes as best you can to its breathless end--if any end there be.[1]
>
> --Woodrow Wilson

"The paradox of an interdependent...economy," Robert Gilpin observes, "is that it creates sources of insecurity and competition."[2] As the regional power program faltered in the mid-'70s, the traditional power planning establishment--the utilities, BPA, and its direct-service industrial customers--sought to maintain the fabric of interdependence through congressional legislation. The utilities needed legislation providing for substantial changes in the framework of regional power planning. The debate created by the legislative proposals drafted by the utilities and others, however, gave vent to the insecurity and competition latent in the cooperation and acquiescence that had prevailed since the imposition of the partnership policy in the 1950s.

The evolution of legislation, starting in 1975, was accordingly both cause and reflection of the widening public debate. Congress, opinion leaders (including state and local government officials), and antagonists (including citizen group opponents and utilities) groped their way toward a workable legislative framework. Oddly but significantly, a severe drought in the 1976-77 water year only temporarily roused mass concern. Although the drought dramatically illustrated the economic perils of electric power shortages, a regional power bill proved to be too esoteric an issue to serve as a continuing focus of public concern.

If the grass-roots constituents were only fitfully active, the technocrats labored mightily to maintain their quiescence. Beginning with Bernard Goldhammer's Round 3 proposal in 1975 and the 1976 BPA Study Paper, legislative revisions of the Bonneville Project Act were proposed that would, it was claimed, benefit everyone in the Northwest. But these claims were at first not understood, then swamped by rising political controversy.

By 1977, when the first major legislative draft was being developed, state and local government officials --long bystanders in regional power planning--had gotten into the act. Oregon led the way in the summer of 1977. Prodded by Governor Bob Straub, the state legislature declared Oregon's intention to take matters into its own hands if the federal government would not solve the state's key electric power problem--the fact that most Oregonians pay higher rates for residential power than their neighbors in Washington. If these rate disparities were not substantially equalized by March 1979, the legislature said, the governor could create the Oregon Domestic and Rural Power Authority--known colloquially as "derpa."[3] DRPA would declare itself a public utility, and claim a share of low-cost federal hydropower on behalf of the domestic and rural ratepayers of the state. The fight for federal hydro would be on, a fight that could destroy the fabric of interdependence. DRPA loomed like a thundercloud over the negotiations for congressional legislation, and gained entree to those discussions for Oregon officials.

The most comprehensive legislative proposal was that developed by the Pacific Northwest Utilities Conference Committee and introduced by Senator Henry

Jackson in September 1977.[4] The PNUCC ("Pea-nuck") bill was a complex, highly technical proposal. It became caught, like much other legislative business, in the maelstrom of the Carter Administration's struggle to enact a national energy policy. The PNUCC bill languished for almost a year, eventually being superseded in August 1978 by a simpler, though still controversial legislative proposal authored by Senator Jackson's staff. The Jackson bill languished in committee as the Ninety-fifth Congress came to a conclusion, but the bill was reintroduced early in 1979 as a "legislative vehicle."[5] As this book went to press in late summer, the prospects for legislation seemed brighter--the Jackson bill had passed the Senate handily. But the possibilities for legislative resolution of the regional power controversy had receded.

Because of its comprehensive scope, the PNUCC proposal provides a useful framework for identifying the issues, actors, and policy ideas surrounding regional power legislation. How that set of issues and actors grew into a public debate is the subject of this chapter. The fragmenting force of that debate will be traced through the evolution of the Jackson bill.

THE ISSUES

Six issues lie at the center of the regional power debate: (1) when (and even whether) new additions are needed to the regional electric power supply; (2) how such facilities will be financed; (3) how the costs of power generation will be distributed among the various classes of wholesale and retail customers; (4) how energy conservation will be implemented; (5) how to plan for regional generating reserves, and the role of the direct-service industrial customers with respect to them; and (6) how to manage the foregoing with revised or new administrative and governmental arrangements. Although the six issues are intertwined, two are principally technological--scheduling of new resources, and planning of reserves. Two others are primarily economic --financing and allocation. And management and conservation are primarily institutional in character. Our purpose here is to outline the enduring contours of these issues, contours which seem likely to shape institutional revision and the continuing problems that those revised institutions must cope with.

Enlarging the Regional Power Supply

The ability of the Columbia River hydropower system to produce more electrical energy has reached economic and environmental limits. To supplement the hydro resource, thirteen thermal power projects--nine nuclear and four coal installations, several with more than one power plant in the project--straggle toward completion, mostly in the 1980s. (See Table 12.) With the ten or twelve years now required to plan, design, site, and construct a major thermal power plant project, generating stations beyond the thirteen projects underway can contribute to the regional power supply system by the 1990s at the earliest. As a result, the accuracy and credibility of electrical demand forecasts have become a central issue.

What amounts to an official forecast is published each year by the Pacific Northwest Utilities Conference Committee. Known as the West Group Forecast, this twenty-year analysis of loads and resources in the West Group area has traditionally projected rapid growth in electrical loads, together with equally rapid increases in the thermal power supply.[6] The PNUCC projections went awry in the early 1970s, as warm winters, rising prices and the economic recession triggered by the 1973-74 Arab oil embargo slowed electrical demand growth.

Utility executives, like others in the business community, expect electric power demand to spurt forward again with economic growth. The West Group Forecast accordingly projects demand growth, although at rates lower than those used in the pre-embargo period. With the slowdown in thermal plant construction--resulting from delays in nuclear plant construction programs--electric energy shortages are projected for the period from the late 1970s through the end of the 1980s, when the five plants of the Washington Public Power Supply System (WPPSS) will have been completed. These shortage projections are based, however, upon drought conditions which would drastically curtail the availability of hydropower; severe droughts did, of course, occur in 1972-73 and 1976-77.

Table 12. *Pacific Northwest Thermal Plant Ownership and Sponsorship.*

Plant		Type	Capacity (MW)	Scheduled Completion[e]	Ownership (%) Private	Public	% Net-Billed	Principal Sponsor[d]
Hanford		Nuclear	860	on line, 11/66	50	50	0	WPPSS
Centralia		Coal	1400	on line, 8/71	72	28	0	PP&L
Trojan		Nuclear	1130	on line, 12/75	70	30	30	PGE
Colstrip	1	Coal	165[a]	on line, 11/75	100		-	PSP&L
	2	Coal	165	on line, 8/76	100		-	PSP&L
	3	Coal	420[f]	July 1983	100		-	PSP&L
	4	Coal	420	May 1984	100		-	PSP&L
Jim Bridger	2	Coal	500	on line, 12/75	100		-	PP&L
	3	Coal	500	on line, 10/76	100		-	PP&L
	4	Coal	333[b]	December 1979	100		-	PP&L
WPPSS Nuclear Project	2	Nuclear	1100	September 1981		100	100	WPPSS
	1	Nuclear	1250	December 1983		100	100	WPPSS
	3	Nuclear	1240	December 1984	30	70	70	WPPSS
	4	Nuclear	1250	June 1985		100	0	WPPSS
	5	Nuclear	1250	June 1986	10	90	0	WPPSS
Boardman (Carty)		Coal	477[c]	July 1980	90	10	0	PGE
Pebble Springs	1	Nuclear	1260	March 1987	85	15	0	PGE
	2	Nuclear	1260	April 1989	85	15	0	PGE
Skagit	1	Nuclear	1288	September 1986	100			PSP&L
	2	Nuclear	1288	September 1988	100			PSP&L

a. Colstrip 1 and 2 each rated at 330 MW; 50% used by West Group Area.
b. Rated at 500 MW; 66% to be used by West Group Area.
c. Rated at 530 MW, of which 90% is to be used by the West Group Area.
d. PGE = Portlant General Electric; PSP&L = Puget Sound Power and Light; PP&L = Pacific Power and Light; WPPSS = Washington Public Power Supply System.
e. As of late 1978.
f. Rated at 700 MW; 60% to be used by West Group Area.

Sources: BPA Role EIS, Appendix A, pp. 1-25, 1-39, 1-40; Part 1, p. 11-15; Pacific Northwest River Basin Commission, Review of Power Planning in the Pacific Northwest-Calendar Year 1976, p. 24; BPA *1978 Annual Report*, p. 10.

Under the strictures of the preference policy, growth in energy demand has affected investor-owned utilities first, and now threatens the power supply of the direct-service industrial customers. Growth in the loads of publicly owned utilities squeezed out firm power commitments to the IOUs in 1973, and--if nothing else happens--are expected to make it impossible to renew DSI contracts as they expire in the 1980s and early 1990s. Indeed, by the 1980s, the growth projected in the West Group Forecast would even endanger BPA's ability to meet public system loads. Thus in 1976 Bonneville administrator Don Hodel issued Notices of Insufficiency to BPA's preference customers. In accordand with the advance notification provisions of the utilities' "requirements" contracts, BPA announced it could not assure provision of firm hydropower to meet public utility load growth beyond 1983. Bonneville's historic mission, to provide electric energy in support of economic growth, is scheduled to come to an end.

Continuation of load growth at historical levels seems out of the question given the sharply rising costs of electric power. The PNUCC forecast, one of the highest for the region, projects growth at 4.5 percent, a pace substantially slower than the 6.9 percent experienced in the period 1950-74.[7] Yet even these rates of expansion pose challenges unprecedented in the Northwest utilities industry.

The need for power has come under close scrutiny, and the West Group Forecast has come under sharp attack from a variety of sources. The most striking challenge is the NRDC Alternative Scenario, assembled by environmentalists in early 1977.[8] Methodologically, the NRDC scenario abandons projections altogether, seeking instead to analyze plausible changes in end use which could be instituted by vigorous conservation programs. By the mid-'90s, even in the face of considerable economic and population growth, the NRDC scenario projects a regional load less than 45 percent as large as that called for in the West Group Forecast.[9] Other analysts, including a group at Washington State University,[10] the Oregon Department of Energy,[11] and the Northwest Energy Policy Project,[12] have also projected lower demand than the PNUCC, using a variety of estimating procedures. (See Figure 16).

Figure 16. *Electrical Energy Load Forecasts for the Pacific Northwest.*

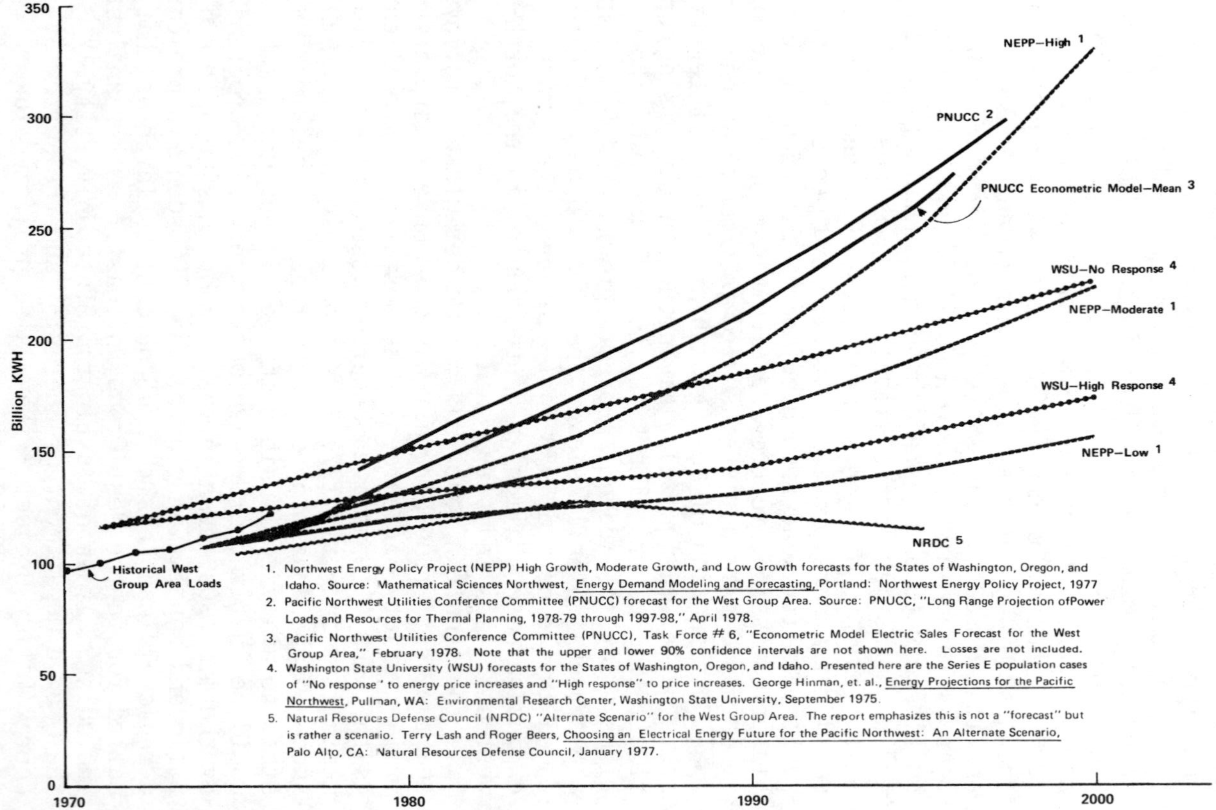

1. Northwest Energy Policy Project (NEPP) High Growth, Moderate Growth, and Low Growth forecasts for the States of Washington, Oregon, and Idaho. Source: Mathematical Sciences Northwest, Energy Demand Modeling and Forecasting, Portland: Northwest Energy Policy Project, 1977
2. Pacific Northwest Utilities Conference Committee (PNUCC) forecast for the West Group Area. Source: PNUCC, "Long Range Projection ofPower Loads and Resources for Thermal Planning, 1978-79 through 1997-98," April 1978.
3. Pacific Northwest Utilities Conference Committee (PNUCC), Task Force # 6, "Econometric Model Electric Sales Forecast for the West Group Area," February 1978. Note that the upper and lower 90% confidence intervals are not shown here. Losses are not included.
4. Washington State University (WSU) forecasts for the States of Washington, Oregon, and Idaho. Presented here are the Series E population cases of "No response" to energy price increases and "High response" to price increases. George Hinman, et. al., Energy Projections for the Pacific Northwest, Pullman, WA: Environmental Research Center, Washington State University, September 1975.
5. Natural Resoruces Defense Council (NRDC) "Alternate Scenario" for the West Group Area. The report emphasizes this is not a "forecast" but is rather a scenario. Terry Lash and Roger Beers, Choosing an Electrical Energy Future for the Pacific Northwest: An Alternate Scenario, Palo Alto, CA: Natural Resources Defense Council, January 1977.

This plethora of projections indicates considerable uncertainty about how near-term policy decisions can influence social and economic forces, which are themselves imperfectly understood. In addition, forecasts have become weapons in the political battle, as Seattle's Energy 1990 debate demonstrated. Nonetheless, there is broad agreement that a regional forecast suitable for planning purposes needs to be developed. There is, unfortunately, no clear way to make such a forecast objectively.

Additions to the power supply can take many forms. At present conventional thermal generation using coal and nuclear fuel dominates the scene. Yet conventional thermal has encountered serious problems. All five of the nuclear plants being built by WPPSS are beset by cost overruns and expensive delays; the Skagit and Pebble Springs nuclear projects have been mired in environmental litigation; and three coal plants in the Centralia and Colstrip projects have been suspended or cancelled for want of air pollution permits. There has been a good deal of enthusiasm developed for less conventional means of generation, notably windpower, which seems to be the best understood solar resource available for Northwest electricity generation.[13] In the meantime, additional hydropower development, raising environmental controversies of its own, is the only supply of any significance which is surely available by the 1990s--and hydro is probably not available in the amounts needed to meet the whole region's needs at moderate growth rates. The unconventional sources may begin to be implemented within the next decade and a half, in time to compete with new thermal. But conventional utility planning calls for rapid growth, requiring hundreds of megawatts of generation; it is hard to be optimistic about new technological options untried at this industrial scale.

Financing

Any thermal project scheduled to go into service in the 1990s must be initiated within the next several years. Given the magnitude of these undertakings, setting up financial arrangements to pay for them is imperative in the near term. Four aspects of financing have been most troublesome: the role of the federal government; regional sharing of the risks of thermal plant development; tax exemptions; and sharing of costs.

The central issue is the federal role. Throughout the hydroelectric period, as well as in the first phase of the Hydro-Thermal Program, the federal government, through the Bureau of Reclamation and Army Corps of Engineers, has built most of the Northwest regional power supply. A major exception has been the "partnership" arrangements by which public utility districts along the mid-Columbia built dams financed by long-term sales to IOUs. With rising thermal power costs, the net-billing capability of Bonneville has proved inadequate to finance additional thermal projects. Given the traditional wariness of IOUs toward federal construction of power plants,[14] the least obtrusive form of federal financing would be for Congress to authorize Bonneville to make long-term purchases of power from new plants built by non-federal sponsors in the region. BPA purchase authority has become the linchpin of legislative proposals backed by the utilities.

The proposed federal purchase would provide, through federal contracts, payments to the plant sponsor whether or not the plant produced any power; no performance standards have been proposed. In the eyes of investors, a national governmental guarantee to pay for the plant is a substitute for--indeed it is superior to--confidence that the plant's sponsors are technically competent.

From the utilities' standpoint, BPA purchase authority is not a bail-out protecting their investment. The utilities argue that regional electric power investment has been dominated by the development of the federal dams. Most of the region's utilities, especially the publics, have accumulated little equity in their assets, since their principal productive facilities were owned by the federal government. Accordingly, federal purchase is seen as a transitional mechanism, permitting the development of equity at a time when borrowing to build high-cost thermal plants might be difficult, given the small amount of collateral available to the utilities. The principal cost of the "equity problem" is, however, higher interest rates, not an inability to provide financing.[15] In a period of high uncertainty over construction costs and the cost of capital, the estimation of costs attributable to insufficient equity alone is impossible.

As for the guarantees to pay back investors whether or not a thermal plant operates, the utilities argue that Bonneville would simply be facilitating regional sharing of risks in new plant investment. The U.S. Treasury, and taxpayers outside the region, will never be asked to bail out Northwest utilities according to this argument, since Bonneville will simply raise wholesale rates to cover the costs of inferior performance, should it occur. Indeed, part of the planned 1979 wholesale rate increase for BPA hydro results from costs incurred by the unplanned shutdown at the Trojan Nuclear Project, and the delays encountered by the three net-billed Washington Public Power Supply System plants. Regional risk sharing is also used to justify legislative language exempting the region's utilities from provisions of antitrust law and the Holding Company Act. These would facilitate cooperative planning, especially by investor-owned utilities.

Oddly, the financing which would be provided by BPA purchase authority denies one of the advantages enjoyed by the public utilities. Public utilities can issue bonds bearing interest exempt from federal income tax. This leads to a lower market interest rate--typically two percentage points lower than that charged corporate borrowers such as IOUs. If the funds raised by such tax-exempt bonds are used to construct a power plant, however, no more than 25 percent of the output of that plant may be sold to a "non-exempt person," under the provisions of the Revenue and Expenditure Act of 1968. The federal government is a non-exempt person, so a specific exemption is needed in the law if the financing advantages of the publics are to be preserved.

An issue implicit in the financing of regional power planning is the relative contributions to be made by public systems, IOUs, and direct-service industrial customers. Under the preference policy, the IOUs have so far borne a major share of the financing burden, notably in the coal-fired plants which serve their loads. The publics have also become deeply involved through their support of the five WPPSS nuclear plants. Much of the competition for low-cost federal hydro is linked to the question of how much individual utilities would have to put up in the financing of facilities to meet the load that hydro cannot cover.

Allocation of Costs

After the need for new facilities and a regional mechanism for financing their construction has been settled, there remains the problem of dividing up the costs of providing electric power to the region's customers. Here the "preference and priority" accorded public systems has skewed the discussion of how costs for the total hydro and thermal systems should be paid. Attention has been focused on the allocation of federal hydropower, or what is the same, allocation of the economic benefits of low-cost federal hydro.

Denied firm federal hydro resources, the investor-owned utilities have faced sharply increasing rates.[16] The completion of the Trojan nuclear plant in 1975 has quadrupled utility rates in Portland within three years.[17] Disparities between private and public system retail rates have made dramatically tangible the inequality of access to federal power.

It should be noted, nonetheless, that the retail rates charged by a utility system depend on a number of factors, of which wholesale power cost is only one. Distribution costs vary widely, with rural systems having especially high costs of service. Each investor-owned utility, moreover, has a distinctive pattern of investments and contractual arrangements, so that its overall cost of wholesale power reflects past decisions. For example, Puget Sound Power and Light has purchased substantial amounts of power on long-term contract from the mid-Columbia PUDs which developed dams in the 1950s. In fiscal 1977, 81 percent of the electricity sold by Puget Power was produced by hydroelectric sources,[18] and more than two-thirds of that came through long-term contracts.

These complications have been largely ignored in the public debate. Instead, attention has focused on the possibility of reinterpreting the preference policy. The argument has been made, primarily by Oregon officials, that the language of the Bonneville Project Act should be interpreted to give preference to a specific class of end users, "domestic and rural customers," rather than to a particular class of distributing utilities, "public bodies and cooperatives." However controversial such an interpretation, its populist premise makes eminent sense in Oregon, where many residential and rural consumers are served by investor-owned utilities denied firm federal hydro.[19] But if domestic and rural consumers are to receive the benefits of federal hydro, these benefits must come from the pool of resources now designated for public utilities, or there must be a subsidy from another class of user, such as the direct-service industries. The "domestic and rural" language has become the focus of a broad-ranging political battle.

A related allocation issue of growing significance is the division of the economic benefits from the sale of secondary power. Secondary power is electricity produced from water resources that cannot be counted on from one year to the next. The amounts of secondary power generated and sold vary widely, therefore.[20] Secondary power is increasingly valuable despite its intermittent availability, since it can be used to displace other sources of electricity, all of which are much more expensive than federal hydro. In 1975, utilities in the Southwest purchased over 1,000 average megawatts of secondary power from BPA, at a cost of $28.7 million; they saved about $167 million in fuel costs alone.[21] The cost differential between secondary hydro and other supplies seems likely to grow larger: one estimate is that the comparative value of secondary power will double by the end of the 1980s.[22] Today, secondary revenues are used to offset wholesale power rates charged Bonneville's wholesale customers; BPA estimates that wholesale firm power revenues would have to rise 25 percent if secondary sales were eliminated.[23] The elimination of secondary revenues is not in prospect--but their benefits might be diverted away from BPA's wholesale customers, as the PNUCC bill proposed. The economic stakes here are sizable: in fiscal 1977 preference customer sales brought in revenues of nearly $125 million,[24] and a quarter of that is more than $30 million per year.

In some approaches, the conflict between the allocation of costs and the preference policy is sidestepped by establishing different pools of power. Each utility or group of consumers is then permitted to draw from one or more pools under specified rules. Much of the discussion of legislation has focused on the details of rate-pool design, especially the consistency of rate-pool structures with the existing preference policy.

Conservation

There is a consensus, established in the course of the regional power debate, that "cost-effective" conservation on a regional scale is needed. While the meaning of "cost-effective" remains to be exactly defined, there is broad agreement that the level of conservation investment should be gauged against the savings from avoided investment in generation. The position taken by Seattle officials, that conservation is cost-effective when it achieves savings at or below the total delivered marginal cost of power, is perhaps the most comprehensive.[25] Thus, if a utility's growth requires additional investments in generation, transmission, and distribution costing 40 mills/kwh, then conservation investments capable of reducing consumption at a cost lower than 40 mills/kwh should be required, and financed by utilities or government authorities.

The agreement on cost-effective conservation represents substantial compromise between utilities and environmentalists. As with any compromise, considerable difference remains on the interpretation of the agreed-upon policy. To the utilities, conservation is imperative as a short-term solution to the shortages projected during the next five to seven years. Given the delays in the thermal projects underway, together with a continued rise in demand, conservation provides the only means of avoiding shortages, especially in IOU service areas, if drought conditions should occur.

To environmentalists and some government officials, however, the chief benefits of cost-effective conservation lie in long-run reform. The cost-effectiveness test, if backed up by tough regulatory standards, can, from this point of view, lead consumers and utilities to undertake conservation measures similar to those they would invest in if electric power were priced at marginal cost. This could be done, however, at prices far below marginal cost--with regulations whose social burden may be substantially lower than the imposition of high electric power prices. Once cost-effective conservation policies are in place, moreover, residential and business users may be susceptible to reforms in energy use with even longer term payoffs than those implicit in present-day marginal costs of production.

Despite the consensus, institutional barriers hinder implementation of conservation programs by utilities acting alone.[26] Public utility financing of conservation investments like building insulation has been hampered by state constitution restrictions on the lending of "the public credit" to private interests--including individual home owners. Although state attorneys general are reexamining the interpretation of this language, constitutional amendments may be needed, in some cases at least, to authorize public utility loans to consumers and businesses for conservation purposes.

Investor-owned utilities are regulated by state utilities commissions, and thus must seek permission to undertake conservation investments. The commissions must approve conservation programs and, perhaps more important, they determine whether utility investment in conservation can be included in the "rate base." The rate base comprises those investments which the regulatory commission counts as productive capital assets, and on which the utility earns a regulated rate of return. The return on the rate base is the principal determinant of the revenues collected through electric power rates. Generating facilities have traditionally been incorporated in the rate base, but state utility commissions have been reluctant to add conservation investments automatically, because of the difficulty of distinguishing between levels of conservation that are cost-effective and so-called "gold plating" investments made merely to increase the return to the utility. Under prodding by IOUs in both Oregon and Washington, nonetheless, utility commissions have been permitting experimental utility-financed conservation programs, with investments included in the rate base.

A regional approach to conservation investments may provide a useful alternative to the utilities' efforts. In the Northwest's integrated system conservation is a regional problem. Moreover, conservation is basically a regulatory policy, requiring governmental authority to be exercised. Most utility service areas --especially those of privately owned utilities--cut across jurisdictional lines. As a result, conservation policies will be patchwork at best. More broadly, regionwide incentives for conservation could provide a means of allocating the benefits of conservation. In this sense, regionally financed conservation is the counterpart to the financing of new generation and the allocation of federal hydropower. In principle, there are two approaches. A regional authority could impose penalties for failure to conserve, or rewards, probably economic in form, could be given for successful conservation policies.

The legislative language drafted by the PNUCC in 1977 proposed that penalties be levied against jurisdictions that failed to institute conservation standards which had been agreed upon by the governors of the four Northwest states. The penalty would take the form of a 50 percent surcharge on wholesale power rates charged by Bonneville. This proposal illustrates the difficulties of region-scale policy making. First, the fact that the police powers needed to implement conservation lie in the province of state government requires unanimity of purpose, and coordination among the states. Haggling over definitions of cost effectiveness could delay agreement on conservation standards for a considerable time. Second, the test of compliance is the instituting of standards, not their enforcement. Third, the use of wholesale rate surcharges as a tool of economic inducement could well prove too blunt an instrument given the highly fragmented patchwork of utility service areas.

Providing rewards for successful conservation policy turns out to be almost as vexing. Consider the case of Seattle. The city government has estimated that its Energy 1990 conservation policy will net a savings to the Pacific Northwest region of $123 million, the cost of the additional facilities which will not be required to meet load growth in the Seattle City Light service area. Of this total savings only about 40 percent would accrue to Seattle in the form of lower rates under the PNUCC legislation.[27] More generally, the design of cost allocations can penalize utilities which succeed in reducing their loads, or in holding them fixed. This was the case in the way that public system access to federal power was defined in the PNUCC bill.[28]

The need for regional coordination of conservation policies remains, since utility service areas grow at different rates. As the economic and technological interdependence brought by thermal generation plays a larger and larger role, the apportionment of new generation costs and the savings from conservation will become economically important. Region-scale coordination is also required to deal with migration into the region, one of the dominant causes of rapid growth. Load growth caused by migration imposes costs on the region as a whole in the form of additional generating resources. Yet, under existing arrangements, public utilities "export" the costs of their own load growth by enlarging their claims on federal hydro. Some redistribution of costs and benefits should be carried out at the regional level; the urgency of doing so is underscored by the rate disparities.

Not only is regional coordination of conservation regulations needed, but the financing of conservation seems likely to need regional reinforcement as well. Cost-effective conservation is only cost effective over a long period of time. As a result, financing is needed, and this may be beyond the capabilities of utilities, especially at a time when their financial resources are being strained by the need to build new generation. Moreover, the constitutional restrictions on public utility loans, and the difficulties of concerting private utility action with that of local governments, could both be leapfrogged by regional conservation financing arrangements. The PNUCC proposal included a $300 million revolving fund to be administered by BPA.

Reserves and the DSIs

The technological change underway in the Pacific Northwest is altering the historical position of the direct-service industrial customers with respect to the regional power supply system. When they came to the Northwest originally during the Second World War, the DSIs purchased power for which there was no alternative market. As DSI spokesmen point out today, the availability of power was an important motivation to locate in the region.[29] As the region's demand has grown, markets for power used by the DSIs have grown too. Now DSI access to federal hydro is based upon the ability of electroprocess facilities to use power that may not be available all the time. To understand the changing position of the DSIs, the function of reserves must be examined in some detail. (See Chapter II for a description of the DSI load and contractual relationship with Bonneville.)

To assure a reliable supply of electricity to meet demand, a utility system must provide for shortages--extended periods of insufficient supply caused by plant delays or unanticipated load growth--and outages--short-term interruptions, usually caused by extremely cold weather or equipment failure. The size of a power supply system is characterized by two measures, its *peak* capability, and the somewhat lower *average* rate at which energy is delivered. Reserves must be planned for both shortages and outages, sufficient to assure both average and peak system capability. The provision of reserves is technically complex.[30]

Hydroelectric generation is subject to unreliability from fluctuations in water supply. Moreover, the hydropower system was built under conditions in which the average supply of power was consistently larger than the demand, often by substantial amounts; water was simply spilled over dams when there was no market for additional power production. Historically, then, reserves were needed only for shortfalls of *supply*. Today, when the costs of building new generation are large, and their financing bears directly on the health of individual utilities, a "reserve" function is also served by having backup *demand*, to absorb supplies which would be temporarily surplus, but which must still be paid for. In this sense, the scope of the reserve function has expanded, and with it, the role played by large industrial consumers of power.

The IF-1 contracts which govern direct industrial sales put the burden of the planning-related delays on the DSIs, by designating one of the quartiles of power as plant-delay reserves. In addition, the industrial customers have signed contracts with the Washington

Public Power Supply System to purchase power from WPPSS Nuclear Projects 4 and 5 in the early years of plant operation, before the participating utilities expect to use all the power from these plants to supply their own loads. Thus, the DSIs provide a market not only for the reserves of federally sponsored hydro-thermal systems, but a cushion as well for some of the thermal facilities which do not enjoy federal backing. These "reserve" functions of the DSIs are served in parallel with their traditional role of providing hydro system reserves through the purchase of interruptible power.

A central issue in continued access to federal hydro for the DSIs is whether utilities would purchase reserves now sold to the industrial customers. Three of the four quartiles of IF-1 power purchases by the DSIs are counted as part of the *firm* load of the region in planning for new generation resources. Yet the IF-1 contract provides for only one quartile to be sold to the DSIs on a firm basis. The other three quartiles provide the shortage and outage reserves discussed above. Outage reserves have a limited market, since purchasers of this power must be able to shed load, or else obtain alternative generation instantaneously. However, shortage reserves, which insure against delay in construction of plants and unexpected growth in load, are rarely used without advance warning of months or even years. Power dedicated to these reserves is marketable to DSIs, utilities, or other wholesale purchasers when the reserves actually exist. If plants are delayed, however, planned reserves can be "on paper" only, unavailable to supply additional loads resulting from drought or unanticipated surges of demand. In sum, the usability of reserves, and their economic value, varies considerably.

It has been estimated that no more than half the DSI load has been curtailed at any given time for any duration.[31] This would indicate that, while only one quartile of power currently sold to DSIs is officially counted as firm, a second quartile--roughly 650 megawatts[32]--may actually be marketable on short-term contracts to utilities in the Northwest or the Southwest. These utilities would have to have sources of replacement energy for periods in which short-term contracts were unavailable from BPA because of demands within the BPA system. Such a capability has traditionally been provided by standby resources such as gas turbines.[33] Such standby resources are costly, but overall production costs are increasing too. The reserve power which could be marketed on short-term contracts should become increasingly attractive to utilities, meaning competition for the DSIs. These reserves could be available for sale to a range of wholesale power buyers--including DSIs but not limited to them. The price charged for these megawatts should reflect their higher quality and broader appeal as short-term firm power, and would thus be considerably higher than the interruptible hydropower rates which have been charged in the past.

Unfortunately, the fact that the need for reserves in the region is changing makes it difficult to project the size of the power resource that will, in practice, be firm. In particular, delays in power plant construction, if combined with rapid growth in load, may lead to the interruption of more than half the DSI load in the near future. From a traditional position as consumer of power which no one else wanted, the DSIs have now become the buyer of resources which no one else can count on.

The reserve function of the DSIs has been widely misunderstood in the region. The technical complexities of DSI contracts have been overshadowed politically by the fact that electroprocess industries consume large quantities of power while providing relatively few jobs. The industrial customers have responded on two fronts. The first is public relations. Advertising, congressional testimony, and a willingness to meet with interested, often skeptical students and others have characterized a posture of openness in explaining the DSI role.

A second DSI response has been to provide assistance with financing new regional generation. In effect, the industrial customers have been willing to accept a high price for an assured supply of power, even interruptible power. In 1978, the DSIs acquiesced in a legislative proposal which would have had them pay the difference between federal hydro rates and the wholesale power costs being borne by the residential consumers of IOUs. The DSIs, that is, were willing to contribute to resolution of the allocation problem by paying sharply higher rates for electric power, in exchange for long-term contracts. Their willingness to subsidize a class of residential consumers thus provided a measure of flexibility in working toward a regional power compromise. Unfortunately, this public spiritedness went unreported and largely unappreciated in the press.

In sum, the changing function of reserves in the regional power supply system has altered the relationship of the direct-service industries to the other consumers in the region. The dimensions of this changed relationship remain shrouded in the complexity of aluminum and power economics, and are obscured by the reluctance of parties undergoing negotiation to disclose all the facts upon which they determine their own bargaining positions. Nonetheless, it is clear that the economic function of the DSIs in the region has been substantial in the past and will be important in the future, a future in which the industries are willing to pay higher power prices.

Governance and Administration

The awesome complexity of regional power issues carries a number of implications. First, the broadening of the agenda of debate has raised many more questions than the participants can answer quickly. For an extended period of time the vortex of discussion and dissension widened, simply because a much larger number of actors was trying to understand the complex stakes involved. Turbulence and confusion have been a precondition, but certainly not a guarantee, of coherent resolution. Second, the ability of public authority to deal politically with the issues raised has been severely hampered. Significantly, of the foregoing five issues, only the most direct--moderating rate disparities--has become an issue with broad popular appeal. Third, any long-term resolution of these issues must rely heavily upon administrative solutions, since the complexity of the issues cannot be accommodated in the political arena.

The traditional basis of governance has rested in the legal responsibility of utilities to serve their consumers.[34] In the era of declining production costs and environmental laissez-faire, this responsibility could be met through expanding power supply. Even then, of course, investor-owned utilities were reluctant to use electricity as an instrument of social development when it was not profitable. The public power movement, born in response to the political judgment that electric power was indeed an instrument of social development, has come to share the view of the IOUs that lowest possible rates and further expansion of electric power supply comprise a satisfactory definition of organizational mission.

The dominance of the utilities, however, has been undermined by the transformation of the social agenda. It must be emphasized that the utilities and Bonneville --though perhaps reorganized under a new charter--remain indispensable; few consumers wish to do without power or even to cut off its growth. Nonetheless, the purposes motivating decisions to expand power supply have become broadly controversial.

At the regional level, and for the purposes of defining federal policy, the central question is the role of Bonneville. BPA control of regional power planning, achieved through its financing of new generation, has been a source of concern both to utilities and to environmental and citizen critics. An extension of BPA's capabilities to finance new generation could impose a regional agenda on managers of local utilities. The regional, technocratic perspective of Bonneville is augmented by its staff capabilities, which generally far outrun those of the utilities, especially the smaller publics. Accordingly, the 1977 PNUCC proposal called for a utility board of overseers, called the Pacific Northwest Electric Planning and Conservation Organization (PNEPCO) which would have the ability to define the construction program for new power plants.

Citizen and environmental critics have generally supported region-scale planning as a substitute for the disjointed pressures for rapid growth originating from the faster growing utility service areas. From their standpoint, the fear is not that BPA will impose a regional program, but that the regional program will reflect traditional utility interests instead of broad regional interest derived from state and local government and citizen groups. To accommodate their views, the 1978 legislative proposal of Senator Jackson created a consumers' council which would articulate those views, as counterweight to a utilities' council serving some of the functions of PNEPCO.

We indicate in Chapter VII a strategy of institutional design which provides for sensible representation of both utility and non-utility interests. Here, suffice it to note that the administrative and technical functions served by BPA are logically integrated in a regional public organization. Although Bonneville should not, in our view, be the only regional public institution with authorities in electric power, BPA in some form remains indispensable.

These six issues carry strikingly different ramifications for different participants in the regional power debate. It is to an understanding of the participants' perceptions and interests that we turn next.

ACTORS AND INTERESTS

Material, organizational, and ideological interests made concrete by the six issues we have discussed shape the positions taken by seven groups of actors: BPA, public utilities, investor-owned utilities, direct-service industries, citizen groups, state and local government officials, and congressional representatives.

In the course of the debate the positions of these groups have changed for a variety of reasons. First, each group's perception of its own interests has shifted, as evolution of the issues and their linkages to one another has progressed. Second, changes in circumstances have changed each actor's perception of its interest; the utilities, in particular, have become advocates of conservation, as plant delays have continued to widen the projected gap between demand and available resources in the 1980s. Third, none of these groups is wholly unified, and the relative influence of contending voices within each camp has shifted, in part because of the dynamics of the regional debate itself, but largely because of events outside the "dance of legislation." Fourth, bargaining involves the forming of alliances for the division of costs or benefits, alliances typically controlled by more than one group of actors; the DSI agreement to subsidize residential consumers of IOUs is a prominent example. These coalitions influence in turn the position of each group of actors in the wider negotiation. What we concentrate on here, accordingly, are the enduring interests of each group, and we have attempted to deemphasize positions taken for merely tactical reasons.

Bonneville

Historically the leader and at times the only member in good standing of the regional planning establishment, BPA evolved during the 1950s into a backstage technocratic facilitator for the region's utilities. When presidential interest in public power died with the partnership policy, Bonneville came to depend upon Congress--for the last twenty years mainly on Washington Senator Henry Jackson--for political support as well as for legislative authorization and annual budgetary appropriation. Because of its insular regional mission BPA has, with one exception, found it easy to remain formally loyal to executive branch policy. The one exception, of course, is the partnership policy, which shifted the agency's mission dramatically, as described in Chapter II. Even then, however, agency resistance to executive control was slight.[35]

The fundamental legislative issue affecting BPA is the granting of purchase authority to finance new generation. This authority, coupled with the agency's existing role in administering the regional system, would thrust upon Bonneville a regional leadership role which it has not exercised since the 1940s. In particular, a regional financing responsibility would give BPA greater authority to review the performance of sponsors of thermal plants, such as the Washington Public Power Supply System.

Within the agency, the dominance of the power management staff would be reinforced by purchase authority. Power management has been in eclipse since the retirement and death of Bernard Goldhammer, the influential power manager who guided Bonneville during the formation of the Hydro-Thermal Program. Long the first among equals in BPA, the division of power management has taken a leading role in pressing for new legislation. Power management staff worked closely with Senator Jackson's aides in drafting bills, and these BPA officials have been unusually visible bureaucrats in public discussions of the regional power bill. Whether the traditional utility orientation of the power management staff--which presently contains Bonneville's conservation office--can accommodate the broader public agenda remains an open question.

Nonetheless, in combination with its traditional leadership of the smaller public utilities, BPA's influence would be preponderant in the region. Its central role seems likely to grow even with mechanisms for limiting the administrator's discretion. Yet, as we argue in Chapters VI and VII, the mechanisms by which BPA's authority is circumscribed will influence significantly the factors brought into the difficult balancing judgments which must be made on behalf of the region, by BPA, the utilities, state and local government, and the affected interests.

Even without resolution of the regional financing question, Bonneville's authority to allocate hydropower under the provisions of existing law brings to the agency a substantial exercise of power, although it is largely regulatory authority, not power to develop new generation.[36] Allocation authority could be used to provide incentives for energy conservation, although the administrative discretion available for enforcement purposes is sharply limited by the preference policy.

Regional leadership by Bonneville implies a distinctive set of managerial and political perspectives. As befits the scale of the technology it administers, Bonneville has an organizational interest in the regional implications of its decisions. So long as the DSIs remain united and continue to base their claim to federal power on their reserve role, they comprise another regional actor with interests paralleling those of BPA to a considerable extent. Bonneville's regional scope naturally includes forecasts, financing arrangements, and regional reserves. In addition, Bonneville appears willing to develop some regulatory powers with respect to conservation policy. Finally, although the agency can take no official position on this, there is clearly a willingness by some within Bonneville to make the agency a publicly visible regional leader once more, as it was during the New Deal.

Publicly Owned Utilities

As creatures of local government, the public systems exhibit much of the managerial and political variability of their parentage. Public system managers have often been parochial in their outlook, and the preference policy, with its valuable dowry of low-cost federal hydro, gives them much to be parochial about. The publics' concern to preserve their hydro allocations reflects as well the public systems' concern about the technological shift which will carry all the utilities in the region beyond hydropower. Symbolized by the Notices of Insufficiency from BPA, this technological shift brings in its train large changes, many of which are novel to utility managers and, worse, uncertain in their implications. Emblematic of these uncertainties are the unnerving cost increases of the WPPSS nuclear projects.

Yet public system managers are more than fearful traditionalists. The present generation of leadership includes veterans of the New Deal, who recall vividly the days when private utilities would not extend transmission lines to rural areas, ignored opportunities to develop hydro resources, and fought public and federal development. Their legacy of public utility commitment to social purposes lingers on.[37] Although some of the rhetoric is mere rationalization of the publics' desire to retain the preference policy, leading public utilities such as Seattle City Light have done much to develop sophisticated positions on governance, conservation, and forecasting. The public preference policy in its traditional form seems to us outmoded, but the possibility of rehabilitating it, as we describe in Chapter VII, seems to us one of the major promises of institutional redesign.

Investor-owned Utilities

The technological shift that worries the publics has engulfed the IOUs. High bulk power costs, adverse financial effects due to unpredictable demand for power, and environmental opposition to new thermal plants have all plagued the region's private utilities. This has led to four major elements of the IOU position.

First, the private companies are anxious to resolve the question of financing. The preference policy has forced the IOUs to develop new resources ahead of the publics. The problems plaguing utilities in the national capital markets have caused Northwest IOUs difficulties in financing thermal plants. Bond ratings have declined into the range between Baa and A, with accompanying higher interest rates.[38] A related problem is that IOU securities have in recent years sold below "book value:"[39] that is, the price of IOU stock has been lower than the proportional value of the company's assets. Under these circumstances, IOUs must raise capital by additional borrowing instead of by issuing stock. Heavy borrowing can further depress bond quality, leading to a downward spiral in stock prices, worsening the utility's plight. Federally assisted financing through BPA purchase is therefore essential to the corporate well-being of Northwest IOUs. A stable financing environment is needed, particularly if it can be combined with an expansion of the capital base available to finance new resources.

Second, the IOUs face short-run resource needs. Conservation is accordingly imperative, as a hedge against expensive power imports or outright shortages in the event of low water in the hydro system. Conservation is especially desirable, of course, if ways can be found to add the investments to the IOU rate base, making them productive economically in the present framework of rate regulation.

Third, with state utility commission regulators facing consumer pressures to hold down rate increases, the IOUs see a valuable opportunity in the reinterpretation of the preference policy so that it favors residential consumers. Such a policy would quiet consumerist pressures from the most numerous, and potentially the most vocal, constituency. At the same time, access to federal hydro would ease the supply crunch that the IOUs may face during the 1980s. In particular, their commercial and industrial consumers would be in a significantly better position if they were no longer competing for power against residential consumers, who are universally favored by curtailment rules. Moreover, the residential preference issue gives IOU executives a way to be publicly and loudly in favor of consumers, a political benefit for an industry subject to extensive regulation.

At the same time, the IOUs have been jealous of their managerial autonomy. As a result, the IOUs have favored governance arrangements which would leave control of investment decisions with the utilities: "We do not want Bonneville to have the dominant voice," declared Robert Short, president of Portland General Electric, at a congressional hearing.[40] Managerial autonomy, from the IOU perspective, is needed if the dual mission of the company is to be met: provision of reliable service to customers, and return on investment to shareholders and investors in utility bonds.[41]

Direct-service Industries

The position of the DSIs is tied to their role as purchasers of regional reserves. A regional reworking of arrangements for power planning is likely to lean heavily on the DSIs' willingness to pay higher costs for assured power supplies in the future. If a regional consensus cannot be fashioned, however, individual electroprocess industries may abandon the DSI consortium which has subordinated individual company interests to a region-wide position.

This possibility has provided the DSIs with bargaining leverage, since 85 percent of the large industrial load lies within or adjacent to public utility service areas.[42] (See Figure 6 in Chapter II.) The possibility that DSIs might apply for hookup to local utilities--most of them preference customers--has added further pressure to find a legislative solution that will accommodate the industry.

Citizen Groups

Citizen interest in Northwest electric power planning has grown steadily over the past several years. Both national groups such as the Sierra Club and local groups such as the Oregon and Washington Environmental Councils, Seattle's Light Brigade and the anti-nuclear Crabshell Alliance have entered the fray through lawsuits, congressional testimony, and lobbying. Their principal congressional spokesman has been Representative Jim Weaver, a Democrat who represents a liberal, environmental, public power constituency centered around Eugene, Oregon. In early 1977 and again in 1979, Weaver introduced bills which stated in legislative language much of the citizen group position. Many of the ideas in Weaver's bill have made their way into language backed by the utilities and his anti-nuclear stance has influenced his colleagues. Yet Weaver and his environmental supporters have been accorded scant recognition by the utility establishment.

The influence of environmental and consumer groups seems to rest on the introduction of new ideas and in a limited ability to veto proposals put forward by the utilities. This combination of innovation and resistance has been utilized to add public-oriented provisions to legislation. As in the Energy 1990 controversy in Seattle, the environmental-consumer perspective resembles that of the Progressive Movement of the early twentieth century. Heavy reliance is placed on public disclosure by utility planners, and by participation of those outside the utility-industry-government establishment in investment decisions.

Greater public participation, activists argue, will result in three substantive improvements to regional power planning. First, construction of new generation will proceed in a more careful manner, reflective of realistic demand projections and sensitive to the environmental and social costs falling on rural areas, such as the coal-bearing plains of Montana. Second, regionally coordinated and financed conservation policy can be implemented--and seriously enforced. Third, the role of the direct-service industrial customers can be circumscribed, with a maximum amount of federal hydro being made available to other consumers. The proposals for bringing about these ends have not reached the degree of precision achieved by the utilities in the PNUCC bill. This is a reflection of the broader ambit of activists' concerns, and the difficulty of arranging substantive results through procedural reforms alone.

State and Local Government

Although the popular base of citizen activism remains numerically small, the rising importance of energy issues has greatly increased state and local government participation in the regional electric power debate. Some agencies, such as the Oregon Department of Energy, have reinforced the voices of citizen group critics of the utility establishment. In others, notably the Washington State Energy Office, official voices have reaffirmed the central position of the utilities in regional power policy. Advocacy of residential preference has been led by politicians in Oregon, with support from the governors of Idaho and Montana.

The governors have wielded the most significant clout. They are politically visible, and the states' regulatory powers are largely under their authority. Moreover, their statewide jurisdictions correspond with those of senators, and the Senate has been taking the lead in legislation. When the governors of Idaho, Montana, Oregon, and Washington proposed a unified position on amending the Jackson bill early in 1979, their suggestions were heeded with alacrity by those drafting the regional power bill.

More generally, the obvious future importance of energy issues has led to the creation of siting councils, energy offices, and staff positions for energy and conservation analysts. The proliferation of official voices has provided legitimate channels for participation in and criticism of utilities' efforts to restructure regional power development. The shift from hydro to thermal has thrust new responsibility on the states' regulatory authorities, and the economic impact of resource development has heightened state and local government interest in utility affairs. The assumptions which underlie demand forecasting are subject to challenge.[43] The issue of who should determine the need for new generation has been argued in state site review councils, along with consideration of what type and size of plant is most appropriate.[44] And conservation, the economic principles to be applied to rate making, and the distribution of large-scale generating plants among the four Northwest states have all been discussed in governmental arenas. What was once utility domain is now opening up to broader public debate, though not all the governmental voices are critical of the utilities by any means. Because there is a good deal of distrust, communication--especially of values and perceptions--among utility managers, consumer and environmental ad-

advocates, and state and local officials remains uneven. This factor contributes to the confusion attending the bargaining process.

Congressional Representatives

The press for regional energy legislation coincided with the immense struggle over the National Energy Act passed in 1978. In part for this reason, Northwest congressmen and senators remained in the background of regional discussions until after legislation had been drafted by the PNUCC-sponsored coalition. The lack of advance spadework has resulted in considerable disunity within the Pacific Northwest congressional delegation, especially in the House of Representatives. While Henry Jackson retained sufficient control in the Senate to bring about passage of legislation, the 1978 session ended with no new power bill. Disunity within the regional delegation has eroded the ability of legislative leaders like Senator Jackson to make deals with congressmen and senators outside the Pacific Northwest.

The PNUCC bill, drafted as an institutional solution to the region's planning problems, became instead an instrument for educating the congressmen representing the region. The academic truth that education is usually the antithesis of action stalled the utilities' bill, as discussion of the necessity for new legislation became ensnarled in debate over the merits of the PNUCC proposal. Although consensus did develop within the delegation that some kind of action was needed, there remain serious disagreements about allocation of federal hydropower, governance, and the role to be played by BPA. In particular, legislators from Montana argued that their state enjoys few of the benefits, but pays many of the costs, of Northwest power development.

The divisions within the Pacific Northwest delegation are said to be harmful to legislation because of the need to present a united front to others in Congress. This is true, especially if the preference policy is to be modified in some respect. Moreover, the need for water and power in the Southwest could be eased considerably with an interregional transfer of resources. And inflationary alarm about rising power costs in the East, together with the drive in Congress and the Department of Energy toward a nationally integrated electric power supply system, could also breach the insularity of the Northwest by legislation.

The strategic position occupied by Senator Jackson as chair of the Committee on Energy and Natural Resources has no parallel in the House. The senior Northwest congressman on the relevant committees is maverick Jim Weaver, ranking member of the Subcommittee on Water and Power. The subcommittee reports to the House Committee on Interior and Insular Affairs chaired by Representative Morris Udall, Democrat of Arizona, whose constituency, lying in the arid lower Colorado Basin, is experiencing rapid growth in population and power demand, a demand which could be met in part by exports from the Pacific Northwest. Moreover, Representative John Dingell, Democrat of Michigan, chairs the Subcommittee on Energy and Power of the House Commerce Committee. Dingell, an ardent advocate of protecting the salmon fishery and a proponent of a national integrated electric power system, must clear any Pacific Northwest regional power bill that passes through the House of Representatives.

With division in the region and the necessity for congressional bargaining at the national level, it is far from clear that a comprehensive solution dealing with all six of the problems discussed earlier in this chapter is feasible. Should legislation fail in the Ninety-sixth Congress, it seems plausible to look for special-interest legislation that would solve parts of the regional power planning problem. For example, the national influence of the aluminum industry could lead to a restricted BPA purchase authorization to provide electric power contracts for the direct-service industrial customers. By removing the DSIs' willingness to contribute financially to the design of a comprehensive regional schema, the flexibility available to work on the five problems besides reserves would be considerably reduced.

In sum, while each major actor sees a need for some sort of change, the motives and directions of redefinition are quite different. The utilities and DSIs want change to stay within traditional utility institutional arrangements. Their efforts are directed at financing new generation and allocating its costs, including the economic benefits of federal hydro. Citizen critics, on the other hand, focus on gaining access to the decision process. Government, state, local and national, is drawn into the debate, often without a clear sense of how more governmental involvement would help. The conflicting objectives within the utilities, and between the utilities, their customers, and their critics have so far eluded successful compromise. But the quest for a stable balance has been evident in the solutions offered by would-be institutional architects.

"ONCE BEGIN THE DANCE OF LEGISLATION..."

Three comprehensive regional power bills have been introduced in Congress. The earliest, the Weaver bill, has never been taken seriously outside the environmental-consumer activist community.[45] The second, the PNUCC bill, introduced in September 1977, provides the most complete statement of the utility industry's program. With the PNUCC bill mired in controversy, Washington's junior senator introduced the Jackson bill in August 1978, an abbreviated but still comprehensive proposal which failed for want of a unified response by the Pacific Northwest delegation in the House as Congress came to a dizzy finale in October 1978. The Jackson bill is now before Congress once more, having passed the Senate in August 1979. Its comprehensive design has been covered over by a series of amendments aimed at preserving the privileges that already belong to one or another interest group. Some form of the Jackson bill may emerge from the Ninety-sixth Congress, but its claim to create a stable basis for regional power development may be uncomfortably tenuous.

As mentioned earlier, we focus here on the PNUCC bill, a complex, highly technical proposal whose major provisions would have rebuilt a framework for regionally coordinated utility investment; reinterpreted--some said abandoned--the public preference policy; and announced a commitment to energy conservation. It thus comprised the most complete set of solutions to the issues discussed at the beginning of this chapter. The institutional design in the bill emerged from an extended process that took up issues that were hardly new at the time. The course of this discussion is relevant to the broad question of how public decisions regarding complex technologies are made--and how they are constrained. Moreover, the fact that the PNUCC bill did not succeed legislatively is a reminder that comprehensive solutions can prove to be politically unpalatable, however carefully assembled.

The PNUCC Bill

The PNUCC proposal is a collection of policy ideas reflecting responses to the six central issues discussed earlier. Its roots lie deep within the institutional evolution of the regional supply system.

From the Hydro-Thermal Power Program, announced in 1968, came three ideas: cooperative development of thermal resources within a framework of hydro-thermal operation of a regionally integrated supply network; a pricing mechanism that blended high cost supplies with cheaper federal hydro; and a federal purchase system that tied the two together. Shorn of the legalisms of the net-billing arrangement, federal purchase has been a reality for a decade in the Northwest. Phase 2 of the Hydro-Thermal Program, announced in 1973, introduced the notion that federal coordination should be subject to limitations exercised by the utility industry.[46]

Legislative action similar to that embodied in the PNUCC bill was proposed first in 1975 by Bernard Goldhammer, who had retired as power manager of BPA, and was working as a consultant to the DSIs. In the "Round 3" proposal released late in 1975, Goldhammer introduced four ideas which have persisted in subsequent legislative drafts.[47] The first was BPA purchase authority, which Goldhammer clearly saw as the basis of a revitalized leadership role for Bonneville. Second, purchase authority should be extended from the existing twenty-year contract period permitted under the Bonneville Project Act to thirty-five years, consistent with the economic lifetimes of thermal plants.

Third, Goldhammer abandoned the utility tradition of uniform wholesale rates for firm power. The region's generating resources would be divided into two blocks. The first block comprised federal hydro and net-billed thermal; the second, thermal additions to the existing system. Power in the second, high-cost block would be sold to all customers, while that in the first would be reserved for preference customers. The preference policy was thus modernized to the extent that net-billed thermal would be included in an averaged low-cost pool of power, while the IOUs would rely on their own resources and regionally constructed thermal power.

Fourth, Goldhammer articulated the DSIs' willingness to recognize their changed position with respect to reserves, and to contribute financially to the technological transition that would affect all consumers in the region. Roughly half the DSI power would, under Round 3, be sold on an interruptible basis on long-term contracts. The other half would be provided on a firm basis, but at increasing cost, as DSIs would gradually purchase increasing shares of thermal power, and release federal hydro and net-billed power for purchase by the preference customers. The use of the DSIs as a financial cushion has proved to be an enduring concept in legislative discussion, as have purchase authority, thirty-five-year contracts, and the idea of dividing the region's power into separate cost pools. Although Goldhammer died in 1976 and Round 3 never gathered much political momentum on its own, his ideas influence the debate today.

In the autumn of 1976 Bonneville issued a "Study Paper for Regional Power Planning," which consolidated --from BPA's point of view--the utility industry view of how the region should proceed. Bonneville would once again lead:

> BPA would purchase output from resources developed by the utilities and other approved entities in amounts adequate to serve the forecasted load growth requirements. Regional conservation programs would be developed and implemented in conjunction with the acquisition of resources.[48]

Not only was conservation mentioned, but "the proposal for development of generating resources would be submitted to the region's States...for comment."[49] The involvement of outsiders was, however, decidedly pro forma, as was a nod to "reasonable incentives" for conservation.

By far the most significant contribution of the Study Paper was its approach to dividing up the costs of power, an approach that quietly breached the preference policy. The Study Paper elaborated Goldhammer's two-pool concept, providing for three pools of power. The first was a hydro and net-billed pool of about 4,000 average megawatts, an amount roughly equal to the thermal power to which public utilities have lent their credit under net billing.[50] This low-cost block would be assigned to the preference customers, its costs frozen at the cost of hydro and net-billed thermal in 1983. A second block of power, composed of the latest addition to the regional system, would provide high-cost thermal generation to IOUs and DSIs denied access to the low-cost preference pool. The third block of power, known as the "BPA Pool," would be made up of the federal power not included in the preference block, approximately 5,500 average megawatts, together with thermal power, drawn from the marginal-cost pool as demand rises. The BPA Pool would be available to industrial customers, investor-owned utilities, and preference customers at a price reflecting the average of federal hydro and the lower cost thermal supplies in the system. The BPA Pool would abandon the preference policy, since it would provide access to federal hydropower to IOUs and DSIs on a continuing basis, even as preference customer demands grow. This design was used in the PNUCC bill, and the weakening of the preference clause ultimately doomed that proposal.

In the fall of 1976 environmental leaders from the region met in Portland to mull over their decidedly ambiguous situation. During the spring and summer of 1976, environmentalists had been invited to participate in the drafting of the BPA Role Environmental Impact Statement, the massive study dictated by the suit which had stalled Phase 2 of the Hydro-Thermal Program. As the summer wore on, the environmentalists' oversight committee concluded that Bonneville was not interested in framing any genuine alternatives in discussing the role of the agency. A request that a conservation-oriented alternative be developed was met by lack of enthusiasm and a refusal to provide BPA administrative support. Led by the Natural Resources Defense Council, the environmentalists resolved to try anyway. Their study, which became the NRDC

Alternative Scenario published early in 1977, sought to map out ways in which the use of energy could be held back by aggressive conservation policies.[51] A major element of their argument was an estimate prepared by the architectural firm of Skidmore, Owings and Merrill for the Role EIS, suggesting that some conservation measures are one-sixth as costly, per kilowatt-hour supplied or avoided, as investment in thermal generation. The environmentalists felt that the argument for cost-effective conservation investment rested on a firm economic base, in a way that the Seattle Energy 1990 debate had suggested might be politically viable.

The environmentalists argued strongly that conservation was financially sound, technically feasible, and most of all, that it would make a big difference. The NRDC Alternative Scenario suggested that thermal power stations then underway could be completed, but that no additional generating resources would be needed in the region until the late 1990s; accordingly, planning for those facilities could be deferred, while policy makers concentrated on conservation policy.

The assumption of conservation leadership by BPA would require new stewards of the public interest at Bonneville. Republican Don Hodel's excoriation of the "prophets of shortage" still rankled. But, in a quiet demonstration of the degree to which congressional influence controlled the affairs of BPA, the environmentalist sympathies of Carter's Interior Secretary Cecil Andrus proved unavailing; Hodel stayed in office, with support from Congressman Lloyd Meeds and the quiet backing of Senator Jackson.

The environmentalists' opposition to Hodel had nonetheless underscored the potentially central role to be played by the Bonneville administrator. Once the PNUCC framework collapsed, attention turned toward building a legislative framework which would permit power to be exercised for the continuing management of a regional power program--and the Jackson bill focused on the powers to be given a strengthened administrator.

The darkest cloud on the horizon, however, did not take final shape until the summer of 1977, just before the PNUCC bill was introduced in Congress. The Oregon state legislature created a formal mechanism to activate a Domestic and Rural Power Authority (DRPA) in

March 1979, if no regional program was enacted by Congress "which the Governor finds will provide equitable costs of power to all Oregon consumers."[52] DRPA would purchase power from BPA and sell it without profit to domestic and rural IOU customers residing in Oregon. DRPA would acquire IOU property only to the extent it must to own such property to qualify as a public preference customer. This probably means that DRPA would contract with IOUs for use of their lines, performance of line maintenance and other utility services, and would purchase power from the IOUs, as necessary, to meet the total requirements of the IOU domestic and rural load.

If DRPA is activated, its claims to low-cost federal hydro will be subjected to court challenge. Should it survive that challenge, however, there would be strong political incentives for Idaho and Montana to follow Oregon's lead. The Jackson bill acknowledged this dynamic by explicitly providing low wholesale prices to all the region's residential consumers, whether served by public or investor-owned utilities. Tactically, the fact that DRPA was being discussed by the state legislature earned Oregon entry into the discussions which led up to the PNUCC bill--the successful breaching of the utility establishment's club rules.

The Bill Emerges

With the Bonneville Study Paper as a starting point, the utilities began in early 1977 to draft regional power legislation. They realized that the burgeoning concern for "domestic and rural customers" possessed a political persuasiveness that they could not ignore. The complex and intertwined interests of utilities, DSIs, and consumers could be trampled beyond recognition if populist fervor for rate reductions got out of hand. Or worse, stalemate might ensue, with further, frightening delay of the regional power program.

Separate task forces were established to develop policies and language for the three linked parts of the PNUCC proposals: federal legislation; power sale contracts; and articles of incorporation and by-laws for a regional planning entity. An expediting committee was formed to coordinate these related efforts. With time pressing, the expediting committee--composed of public utility, IOU, DSI, and BPA representatives--became the driving force of the negotiations. It assumed primary responsibility for the most controversial decisions relating to the allocation of federal power and the trade-offs to be made among BPA's public, IOU, and DSI customers. Despite environmentalist complaints, the expediting committee continued to operate behind closed doors. Its hurried compromises dealt out rough justice to the diverse interests of the region; the utilities could ill afford extended debate. Yielding to pressure from Portland officials, a fourth task force was established to develop a conservation policy to be included in the bill.

Among the public systems only Seattle, Eugene Water and Electric Board (EWEB), and Snohomish County PUD defended their preference status. The once solid front of public power had been almost wholly converted to the need for a power bill, even one that would sacrifice the "sacred cow" of preference. The PNUCC drafters hoped legislation could be introduced in Congress by April first, with congressional committee review completed by May 15, 1977. The negotiation process delayed introduction until September. Like the thermal plants they were trying to facilitate, the building of a new legislative framework mired the utilities in delays, delays that seemed only to elicit more conflict and more delay.[53]

When the PNUCC bill was introduced, its authors had achieved an important technocratic objective: the economic framework of the BPA Study Paper was intact, facilitating cooperative financing of thermal development and sharing the economic benefits of federal hydro. But that technocratic success was purchased at high political cost: resentment over lack of participation by non-utility public and governmental bodies. Worse, the threat to the public preference policy began slowly to surface. The miasma of political discord hung over the bill, and members of the congressional delegation, wary of its taint but unsure of its depth, were cautious even in introducing the PNUCC proposal.

The PNUCC bill remains the most detailed proposal for refashioning the institutional framework of regional power planning. Its provisions thus yield important insights into the utility interests which must be part of any eventual solution. The IOUs and DSIs, who enjoyed unity--and confidentiality of internal debate--seemed to have had their concerns most fully accommodated. The IOU negotiators' goals "were to keep public/private [rate] disparities to a tolerable five mills at the wholesale level."[54] Maintenance of a modified preference for existing public utilities[55] was another constraint, though the modification ultimately proved unacceptable to the publics. The PNUCC expediting committee shared other negotiating assumptions: that long-term BPA purchase of plant output was essential; that a regional conservation program was appropriate; that planning and designation of future power supplies were essential at the regional level, but that this function should be a utility rather than a governmental responsibility. These assumptions were firmly grounded in the belief, projected in the West Group Forecast, of a severe electrical shortage beginning in the mid-1980s if water fell to drought levels.

Each of these negotiating assumptions influenced the design of the three parts of the PNUCC proposal: the bill itself, the "Regional Power Plan"; the articles of incorporation of the Pacific Northwest Energy Planning and Conservation Organization (PNEPCO); and the power sales contracts. The legislation presents the general, overall proposal; the eleven types of contracts and the articles of incorporation detail specifics. The three formed an integral whole, however, and this, too, was to provide some problems in maintaining consensus among utility and governmental entities once the proposal came under attack.

In this setting, the PNUCC bill set forth the utility position on each of the six issues discussed earlier.

1. *Enlarging power supply*. A regional forecast would be published by PNEPCO, following the tradition of the West Group Forecast. PNEPCO would also evaluate proposals for new resources and designate resources eligible to have their output purchased on long-term contracts by Bonneville. PNEPCO would thus schedule power plants for the region. Contractual arrangements, as Goldhammer had suggested, would run for thirty-five years.

2. *Financing.* BPA would, with PNEPCO approval, enter into contractual arrangements for the purchase of the capability--the potential output--of new thermal plants.

3. *Allocation of federal hydro and thermal costs.* Three blocks of power were defined in the bill, together with formulas to determine access to each block. This was basically the structure of the BPA Study Paper; the three pool approach involved a permanent assignment of federal hydro to nonpreference IOU customers. If a plant failed, becoming a "dry hole," its costs would be spread proportionately among wholesale customers in all three rate pools.

4. *Conservation.* A link is forged in the program between conservation on the regional scale and BPA provision of federal power. Up to $300 million would be dedicated to a revolving fund supervised by Bonneville, for conservation grants or loans to be administered by utilities. Conservation standards would be set by unanimous agreement of the four Northwest governors. Utilities located in states which fail to implement the programs would be assessed a 50 percent surcharge on BPA wholesale power, unless BPA determined that the utility was in "substantial compliance."[56] The surcharge could also be waived for any political subdivision found to be in compliance by a unanimous vote of the Northwest governors. Such a commitment of BPA funds to conservation would provide a region-scale alternative to central station generation of power. As noted above, a regional financing mechanism could ease institutional barriers to utility-administered conservation programs in both IOUs and public utilities.

5. *Reserves and the DSIs.* Within the regional power pool, reserves would be available for sale to direct-service industrial customers. This resource would be augmented by an interruptible quartile of outage reserves, which could be curtailed by BPA unilaterally, the whole package being sold to the DSIs under the IF-1 contract. In addition, DSIs would be eligible to purchase power after 1985. New DSIs may be formed; but these new customers would be required to purchase new thermal power, comparable in cost with electric power available elsewhere in the nation.

6. *Governance*. The PNUCC bill, as noted before, is one of three components of a utility industry proposal. Access to the three rate pools would have been controlled through eleven different types of power purchase contracts, the second element of the utility plan. The third element was the articles of incorporation of PNEPCO, a non-profit corporation.

PNEPCO was in essence an official form of the PNUCC, receiving the imprimatur of Congress through the legislation. PNEPCO's membership was limited to utilities, although governmental observers would have been permitted to attend. Public attendance was to be limited.

In the PNUCC plan PNEPCO would be responsible for developing load forecasts and "gating": approving new resources eligible for Bonneville purchase. In this manner, utilities would retain ultimate control over the scheduling of power plant construction. PNEPCO's broad powers raised doubts "whether a trade association is appropriately granted the broad planning authority described by the legislation."[57] But the utilities did not stop there.

The mix of private enterprise and public utilities in PNEPCO required additional exemptions from federal law, for antitrust and tax purposes and to lower financing costs. A tax exemption would have guaranteed the ability of public agencies to issue bonds whose interest is free of federal income tax. Antitrust immunity would have been sought as well, to facilitate utility cooperation in forecasting, price fixing through contractual agreements, and the allocation of power resources. And the IOUs would have received an exemption from the provisions of the Holding Company Act, to facilitate their cooperation with one another.

Finally, the establishment of the conservation program would hinge upon the ability of the four Northwest governors to agree upon conservation standards which they could unanimously support.

In sum, the basic problems of imminent shortage, increasingly disparate rates between public and private utilities, and the need for a firm and relatively inexpensive supply of power to direct-service industries were all to be solved by a new regional program.

It didn't work.

Critiques

The PNUCC plan foundered on two rocks, the preference policy and the allocation of federal hydro. There were loud complaints as well about the lack of public participation in the utility-designed program. The PNUCC bill died in committee, a victim of dissension within the utility industry, compounded by the public complaints of citizen groups and environmentalists.

The bone in the throat of federal hydro allocation was the claim of Oregon's domestic and rural customers. The PNUCC plan had relied on the complex three-pool structure to allay anxiety about rate disparities. Yet even as the PNUCC plan was being put together, the Oregon state legislature passed the law creating the Domestic and Rural Power Authority, with its March 1979 deadline for federal action. A game of political chicken had begun. As the utilities attempted to gather momentum behind their proposal to force DRPA off the road, the City of Portland launched a diversionary attack by filing two lawsuits against BPA late in 1977.

Portland consumers are served by two investor-owned utilities, Pacific Power and Light (PP&L) and Portland General Electric (PGE). In 1974 and 1976 initiatives were voted on and defeated to transfer PP&L's assets to a new municipal utility. During each initiative campaign, Bonneville's legal office had asserted that no firm power was available to serve a new public utility load; this helped to defeat the initiatives. No litigation was initiated, despite a good deal of grumbling about the BPA opinion and its apparent inconsistency with the preference policy.

Undeterred by the two unsuccessful initiative efforts, Public Utility Commissioner Francis J. Ivancie argued for a broad reinterpretation of the preference policy, so that it would benefit domestic and rural customers throughout the Northwest. The Ivancie plan is based on the concept that federal power is a social tool, used in the 1930s to promote rural economic development. But the work to be done with that tool has changed:

> The poverty pockets and the people on retirement and fixed incomes who have been so cruelly hurt by inflation now live, for the most part, in the urban areas. If low cost

> federal power is to continue to be a social tool, then it should be used to help these people, not to promote the economic development of one area in the Northwest at the expense of other areas.[58]

The Ivancie plan would give all residential customers in the Bonneville service area access to preference power. As with DRPA, this power would be sold to residential customers by IOUs without profit.

The claims of the Ivancie plan and DRPA to be the politically coherent successors to public preference proved compelling later on. But in late 1977, in a move that dramatized the lack of consensus behind the PNUCC plan, the City of Portland filed two suits just before the first congressional hearings on the PNUCC bill were held. In one suit, the city argued that federal power should be allocated to domestic and rural customers.[59] The second suit challenged the legality of the utility contracts signed since 1970 on the grounds that they have ignored the National Environmental Policy Act. This litigation challenged the net billing arrangements concluded since 1970. The sale of WPPSS bonds to finance net-billed plants was temporarily stalled by the Portland litigation, although a $180 million offering was sold early in 1978. Both suits were dismissed late in 1978.

As the logic of DRPA and the Portland suits became clear, the governors of the Northwest states took notice. In a series of inconclusive but increasingly tense meetings, governors Robert Straub of Oregon, John Evans of Idaho, and Thomas Judge of Montana, all representing states where IOUs serve a large majority of the residential consumers, lined up against Washington's outspoken Dixy Lee Ray, who stoutly defended the 60 percent of her constituents who are served by public power.[60]

The political momentum for residential preference could probably have been accommodated within the framework of the PNUCC legislation. Since the economic burden of IOU customers would already have been lightened considerably in the rate pool design, a legislative mandate for residential preference would have been relatively easy to add. But as the pressure to residential rates grew, so did the backlash from the public utilities, determined to preserve the inviolability of the preference policy.

The first shot was fired across the bow before the PNUCC bill ever left port. As the PNUCC bill was introduced by Senator Jackson, the Democrat from Everett simultaneously introduced language which came to be known as the Snohomish Amendment. It was drafted by William Hulbert, manager of the Snohomish County Public Utility District, whose service area includes Jackson's home town.

The Snohomish Amendment took a hard line on public preference, recommending a new rate pool comprising the firm federal hydro resources, priced at hydro cost. The 2,700 megawatts of net-billed thermal power would be put into a joint public-IOU thermal pool. The substantial revenues from secondary hydro production would accrue entirely to public utility customers. The other characteristics of the PNUCC allocation proposal, and all other aspects of the bill, were to remain intact.[61] While Hulbert's was on the surface a "friendly" amendment, the blow it dealt the PNUCC bill was fatal: it contravened the basic premise of the PNUCC coalition that rate disparities would be mitigated by using federal hydro to dilute IOU rates.

Moreover, the Snohomish Amendment had attracted the attention of two national trade associations, the American Public Power Association (APPA) and the National Rural Electrical Cooperatives Association (NRECA), long the guardians of public power in federal politics, law, and policy.[62] Early in 1978 the APPA formally adopted a resolution--drafted in August 1977, before the PNUCC bill had been introduced--that backed away from key provisions of the PNUCC package.[63] The Association resolved that it would "not abandon its historic positions with respect to the sale of Federal preference power, the application of the Federal antitrust laws, and the formation of new publicly and cooperatively owned electric utilities."[64]

The facade of public utility support had cracked: by backing away from the PNUCC's revision of preference and its attempt to avoid antitrust restrictions, the APPA told Northwest public agencies their vital interests were in jeopardy. So was the bill.

The PNUCC package had been finely tuned, like a precision timer. But the utilities' inability to contain the advocates of residential preference and the suspicions of public power agencies had caused the timer to malfunction; the package blew up in the industry's faces. Why had the utilities so badly misjudged their own members and natural constituencies? In part the great haste with which the PNUCC package was assembled had led to carelessness. A second, a more fundamental problem was a misreading of the political situation, fostering a sense among public utilities that they had not been fully consulted or fairly dealt with by the PNUCC expediting committee. As the utilities' lawyers worked furiously on the elements of the PNUCC plan, an attorney for the Public Power Council who attended these meetings wrote a troubled memorandum:

> ...Although lawyers can advise, the settlement of this question is beyond their professional competence and responsibility, being rather a policy issue for the responsible policy-making public officials--the commissioners, directors and city councilmen responsible for direction of the preference agencies. These officials' responsibility is complicated by a number of factors: the legislation and contracts are extraordinarily complex, dealing with many situations and relationships that had not been extensively discussed before negotiations began. Few have had the opportunity to evaluate the advantages and disadvantages... [A] situation which heavily loads on that customer the high costs of new thermal generation...should be made very clear to each and every preference customer before we assume that it acquiesces in the new program, for its policy-makers in any event will have a lot of explaining to do when the public served by public agencies begins really to understand...its ...impact on public agency costs.[65]

Not only were the publics left feeling excluded, but so was Portland. Mayor Neil Goldschmidt complained during the negotiations that

> the crucial decisions on power pooling, melding, access to pools, and the composition and functions of the regional planning organization were all made prior to participation by the city and the State of Oregon.[66]

What the Public Power Council lawyer wrote about his clients seemed to stand for the political style which characterized the drafting of the PNUCC program:

> I am particularly concerned as to whether all of the preference agency policy-making officials sufficiently understand the program, including particularly the risks involved, and including particularly the fact that their respective utilities, will be committed as a practical matter once the legislation is enacted. As a practical matter, once that happens there will be no real choice to opt out, at least for most of the utilities.[67]

The Jackson Bill

Don Hodel, who had served as deputy administrator and administrator of Bonneville for nearly a decade, announced his resignation at the end of 1977. In January 1978 he was replaced by Sterling Munro, a long-time aide to Senator Henry Jackson. Democrat Munro at once abandoned the "partnership" passivity of his predecessor and began making speeches throughout the region, cheerleading the development of a political consensus for regional power legislation. As the PNUCC package faltered, Munro tried out a number of statements and positions, seeking to expound a viable regional consensus.

The PNUCC bill had served a valuable educational function for the senators and congressmen themselves. Beginning in December 1977, hearings were held in the Northwest and in Washington on the bill. The field hearings dramatically revealed the lack of agreement within the region, together with the urgent desire by utility managers for a rejuvenated, stable planning framework.

In August 1978, as the embattled National Energy Act neared the final throes of the natural gas deregulation compromise, Senator Henry Jackson called a meeting of the members of his Committee on Energy and Natural Resources from the Pacific Northwest. The senators and the committee staff, presumably buoyed by Munro's news of a viable consensus, rapidly drafted a set of twelve principles on which a new regional power bill should be based. Final agreement on the principles was reached Monday, August 14. On Wednesday morning, August 16, Senator Jackson introduced S.3418, the Jackson bill.

The Jackson bill discarded the finely wrought filigree of the PNUCC program, replacing it with a cast-iron reliance on a strong BPA administrator. The rate pools were dispensed with, and in their place residential preference would be phased in over a five-year period, during which the direct-service industrial customers would, through a series of power exchanges with the IOUs, gradually confer the economic benefits of federal hydro on the residential customers of the IOUs.[68] The scheduling of resource acquisitions was to be guided by a new principle: first, cost-effective conservation would be implemented, then renewable resources would be developed, and last, conventional thermal resources would be purchased. Bonneville purchase authority was thus tied to a sequence of priorities favored by environmentalist critics. The preference clause was not mentioned.

In part because it was introduced with such haste, the Jackson bill encountered a good deal of controversy on its own. Each of the major actors found something to dislike about it. But most important, time was short, as the November mid-term election loomed uncomfortably close for a bill that first saw the light of day in mid-August. Sterling Munro's vision of consensus seemed to attach no more securely to the Jackson bill than it had to its PNUCC predecessor. Although Jackson expressed some confidence that the Senate Energy Committee would consider the bill, and that the full Senate might well have passed it, it was broadly agreed that passage in the House was unlikely.

The bill died with the Ninety-fifth Congress. The utility monopoly on the regional power bill was shattered.

"THIS IS TRIAL AND ERROR"

The Jackson bill was resurrected in 1979, but its legislative purposes had changed subtly. No longer a simple, comprehensive solution to the utilities' problems, the regional power bill was advanced as the best compromise possible. "Let's see how it works," Senator Jackson said of a key provision. "This is trial and error."[69] The reborn bill was amended substantially and passed by the Senate in August 1979; as this book went to press, the House of Representatives was considering its own version of a Northwest power bill.

Proponents of the Jackson bill hoped that a revised version would be signed by the President before the beginning of 1980. To optimists, this law would provide an institutional structure as strong and durable as the Bonneville Project Act. But the bill's language suggested that the entente among the utilities of the Northwest was a fragile one. Though the strong BPA administrator of the original Jackson bill remained, important reservations staked out the retaliatory powers of the utilities, should the administrator's plans falter. The administrator might still be a "regional energy czar," as some had feared, but the duchies and baronies over which he ruled seemed determined to preserve their prerogatives.

A Compromise "Solution"

The Jackson bill as it was passed by the Senate nonetheless did address each of the six central issues of the regional power debate:[70]

1. *Enlarging power supply.* Responsibility for providing adequate generating resources in the Northwest rests with the Bonneville administrator, acting in concert with a regional council on which the administrator serves. As in the original draft, priorities are set forth in legislation: "First to conservation; second, to renewable resources; third, to generating resources utilizing waste heat and generating resources of high fuel conversion efficiency; and fourth, to all other resources."[71] These priorities are to be incorporated in a regional plan and forecast to be developed by a

Pacific Northwest Electric Power Planning Council, whose five members are the Bonneville administrator and appointees of the governors of Idaho, Oregon, Montana and Washington. Projects developed on a regional basis must be "cost-effective and feasible" in the judgment of the Bonneville administrator. The degree of discretion given the administrator by these criteria--especially the undefined term "feasible"--is disputed.

2. *Financing*. Resources consistent with the regional plan may be purchased by the administrator. The purchase authority is defined in terms of "capability" rather than output, making purchase an open-ended commitment as desired by the utilities and opposed by nuclear power skeptics. Financing is facilitated for public power agencies by a tax exemption, and for IOUs by an exemption from the Public Utility Holding Company Act.

3. *Allocation of Costs*. The preference policy is reaffirmed.[72] But, as in the 1978 version, residential and agricultural customers of IOUs are provided the economic benefits of low-cost federal hydro through a power exchange with the direct-service industrial customers.[73] The most important of the rate directives which allocate costs is the so-called "rate test" devised by the Public Power Council after the bill was first introduced. The rate test sets a ceiling on wholesale power rates charged public power agencies, so that they will not have to pay rates higher than they would have paid in the absence of congressional legislation. If the ceiling is exceeded, other wholesale customers of BPA will have to foot the bill; this can happen any time after July 1, 1985. Under these circumstances, however, investor-owned utilities can back out of the power exchange agreements which provide low-cost hydro to their residential customers.[74] If these defections from the regional program occur, the program would be a dead letter--and the regional debate back to square one.

4. *Conservation*. The Jackson bill repeatedly instructs the Bonneville administrator to take the lead in regional conservation efforts. The cost-effectiveness test for conservation is 10 percent less stringent than it is for other resources. Model conservation standards are to be developed as part of the regional plan, and jurisdictions which fail to enact regulations

consistent with these models may be assessed wholesale rate surcharges ranging from 10 to 50 percent. Financing for investments including conservation may be provided through BPA-issued bonds worth up to $1.75 billion. But mindful of the authorities and sensitivities of state and local governments, the bill explicitly prohibits Bonneville from imposing mandatory conservation.[75]

5. *Reserves*. Direct-service industrial customers are expected to turn in their existing contracts for power in exchange for new IF-1 contracts at substantially higher rates. Power to the DSIs is legitimated in terms of their reserve function at the regional level, although Bonneville retains considerable discretion in dealing with the industrial customers.

6. *Management and organization*. The Jackson bill relies on a very strong Bonneville administrator who interprets the regional power plan and may proceed independently of it with congressional sanction. The regional council representing the governors does have substantial clout in the planning procedure, although here too BPA plays a determinative role. Nonetheless, it is probably fair to say, as a Bonneville staff member told us, "The states are crucial."[76]

In three respects the Senate bill is surprisingly strong. First, the commitment to conservation and cost effectiveness goes nearly as far as the most outspoken environmental critics would wish. Cost effectiveness, in particular, is defined in terms of the incremental system cost to the consumer. Moreover, cost effectiveness must include "quantifiable environmental social costs and benefits."[77] Second, despite strenuous lobbying by the region's utilities, no thermal projects beyond the existing net-billed facilities are grandfathered into the regional resources plan. While thermal projects are by no means excluded, efforts of the public power agencies to stack the deck in favor of WPPSS Nuclear Projects 4 and 5 proved unavailing. Third, there is language directing fisheries protection, a dimension of environmental policy which many had thought would be left to the House to attend to.[78] (There is support in the House for stronger language on fishery protection.)

There were still criticisms of the Jackson bill, especially among environmentalists, who saw in the bill a mandate for massive thermal plant purchases. The regional plan, defined as "a broad policy document" by the Senate, need not identify specific generating stations.[79] This leaves much discretion to the administrator of BPA in making resource acquisition decisions.[80] Moreover, although the administrator must submit to Congress an extensive record of his decision to acquire a major resource, the range of judicial review accorded that decision is rather narrow.[81]

Most of all, the "extraordinary latitude granted" the BPA administrator encountered strong resistance from those who felt that BPA will merely endorse the plans of the utilities.[82] Can any administrator, even Sterling Munro, be trusted? The lingering distrust of Bonneville reflected a growing sense of frustration among the pro-conservation forces in the region, who feared that the fix was on in Congress. With anxiety over the nation's drifting energy situation at fever pitch in the swampy Washington summer, Congress had already shown its willingness to bypass environmental review through a national "mobilization board." The Jackson bill contained a worrying ninety-day time limit on suits challenging any "major action" pursuant to the bill.[83] Opponents feared that the urge to "fast track" energy might sweep the bill to passage; it was an unease BPA could only exacerbate as it continued its press for enactment.

These criticisms were sure to be marshalled as the House went to work on the regional power bill. It seemed unlikely, however, that the House would depart from the cautious, incrementalist approach of the Senate. Larger issues might fall through the cracks as each interest group looked out for its own immediate concerns.

What Would the House Do?

The largest factor in favor of the Jackson bill was the support of two key members of the House of Representatives, Representative Al Ullman, Democrat of Oregon, chairman of the powerful Ways and Means Committee, and Representative Tom Foley, Democrat of Washington, chairman of the House Agriculture Committee and a respected legislator. "There is plenty of political clout out there" in the Pacific Northwest, one aide told us, "but it hadn't showed its strength."[84]

The most powerful new element affecting the chances of passage was the Three Mile Island nuclear accident in the spring of 1979. "The bill is going to have to carry an awful lot of baggage on the nuclear issue," Representative Al Swift said."[85] Congressman Jim Weaver had made a big issue of his opposition to nuclear energy and his suspicions of the regional power bill on those grounds. His easy win in the 1978 election showed that his stand was popular at home.

The most important matter of timing, however, has to do with costs. Early in 1980, as this book appears, the wholesale rate increase sought by BPA will appear in consumer's electric bills. "It's an absolute no-win," said Al Swift. "Everybody's rates are going to go up, and they are going to blame it on the bill and they are going to blame it on anybody who voted for the bill. There's no way to duck it."[86] Rising costs will mean more criticism of WPPSS and the authorization of BPA purchase. "You've got a bunch of farmers out there," Congressman Norm Dicks says, "trying to manage a $12 billion a year [*sic*] operation....They are not competent to deal with this thing."[87]

As the design of public policy gives way to the pulling and hauling of legislative politics, there is an inevitable tendency to push hard or complex questions off into the future, in favor of immediate issues. That the utility industry was passing through a difficult period of transition toward a more costly and controversial public setting seemed less important than securing the financial means to resume a building program stalled for nearly five years. If the public power rate test was triggered, and the structure of regional interdependence fell apart, well, that would have to be addressed as it came up.

Legislation had been touted as a comprehensive solution; it is in some danger of becoming little more than the record of interest groups' staked claims. A new bill might be "a way of reordering how we provide electricity," one congressional aide hoped.[88] But a staff member of a crucial subcommittee said instead, "I don't think that the committee is thinking like that."[89] The question now is whether a stable regional structure is feasible at all.

CHAPTER VI
Institutional Analysis

> The essence of great tragedy...is the realization by the self that it is totally unequipped to confront the universe. We might venture that it is even more tragic that [man] is unequal to the task of dealing with a universe of his own manufacture. Gloucester in *King Lear* blames the gods who kill us for their sport, as wanton boys do flies....But whom dare we blame for Gary, Indiana?[1]
>
> --Perry Miller

With or without a new legislative charter for the Bonneville Power Administration, the Northwest electric power system faces important changes in technology, economics, and administration--all to be carried out amid political tension and occasional conflict. Energy has become central to social change and economic development, a role that New Deal visionaries claimed for it four decades ago. Regional power policy has become a way to influence, perhaps decisively, the future of the Pacific Northwest. Yet the link between legislative proposals and alternative regional futures is poorly understood--if powerfully sensed. Thus the conflict over which goals to pursue is complicated by controversy over the means for attaining them.

Choosing the future, if it could be done at all well, would surely rank among the most profound political decisions a society could make. Yet these choices are rarely made explicitly in a democratic society, in part because such a choice can sharpen conflicts latent among its citizenry. The American political order has been based on pragmatic compromises, compromises that promise something for everyone and a clear choice for none. The unusually taxing challenge of Northwest power politics reflects the complexity of the search for compromises, together with the difficulties of representing the spectrum of interests active in the debate. Under these circumstances institutional change must take into account both the historical constraints of Northwest power development and the social and technical opportunities open to the region. In this chapter we attempt to estimate the scope of the political conflict that a revised institutional structure must deal with.

We examine four questions. First, why did the Bonneville Project Act endure as well as it did? Second, why is the act no longer workable? Third, will new legislation help, and if so how? Last, what measures besides new legislation are needed? In taking up these questions we shall follow the sequence of events described in Chapters II-V. (See Table 13 for a chronology.) Our emphasis here, however, is analytic rather than descriptive.

The durability of the Bonneville Project Act highlights the role of unofficial and informal ties linking BPA to its utility clients and Congress. This pattern is one often found in American politics. By the late 1960s, however, the agency-client-Congress framework had been undermined by three forces: the unexpectedly high cost of thermal power; political tensions associated with high thermal costs in the context of the preference policy; and the expansion of public concerns linked to electrical power, including environmental quality and energy conservation. The legislation that seems able now to pass through the congressional gauntlet may resolve problems arising from economic interests, but that may not be enough for the long run. Accordingly, as we shall argue in the next chapter, institutional revitalization of BPA and other organizations active in regional power is imperative if the inevitable conflicts associated with electric power are to be fairly and successfully resolved.

Table 13. *Evolution of Institutional Structure in the Pacific Northwest Electric Power Supply System.*

Date	Political Stimulus or	Technological Development	Institutional Response
1933-46	REGIONAL ECONOMIC STIMULUS		
1933	Public works for economic stimulation	Multiple-use river development	Bonneville and Grand Coulee Dams authorized by Congress.
1937	Public power movement, together with opposition to TVA		Bonneville Project Act, with marketing authority only. Public power grows.
1940		Technically advantageous industrial load	First DSI contracts; economic growth.
1942	War production	Regional interconnection	Northwest Power Pool.
1946-52	INTERREGNUM		
1946		Cooperative information exchange	Pacific Northwest Utilities Conference Committee
	Conservative 80th Congress elected		Slackening of public power growth, and weakening of BPA leadership.
1952-60	PARTNERSHIP		Withdrawal of federal support for new Columbia River dams.
1957-65		Cooperative hydropower development	Non-federal mid-Columbia dams.
1957	Public Power reaction		Washington Public Power Supply System.
1960-73	TECHNOCRATIC AUTONOMY: THE HYDRO-THERMAL CONCEPT		
1962	Jackson Compromise	Dual-purpose nuclear project	Hanford New Production Reactor authorized by Congress.
1964	Columbia River Treaty	Upstream storage	Canadian storage dams and Libby Dam, financed by Columbia Storage Power Exchange.
		Cooperative hydro operations	Coordination Agreement.
		Inter-regional connection	Pacific Intertie authorized by Congress, with regional preference.
1968		Hydro peaking and thermal base load operation	Hydro-Thermal Power Program; Net Billing.
1973	STRUGGLE FOR THE FUTURE		
1973		Rising thermal plant costs	Net billing discontinued.
1974	Environmental opposition to HTPP, resulting in Governors' request to Council on Environmental Quality		
1975	Port of Astoria *v.* Hodel, NRDC *v.* Hodel		BPA Role EIS.
1976	Public-private rate disparities	Energy conservation	Seattle Energy 1990.
			Rate disparity issue and criticisms of DSI role.
1977	Attempt to build utility consensus		PNUCC bill.
			Oregon DRPA; Portland *v.* Hodel *et al.*
1978	Conflict over preference clause and hydro allocation		PNUCC bill falters; Seattle City Council resolution. Jackson bill introduced.
1979		WPPSS attacked for cost increases	Washington State legislature orders audit.
	Public Power Council amendments Governors' amendments		Jackson bill amended.

Our ideas for encouraging such a revitalization are set forth in Chapter VII, and the readers whose primary interest lies in near-term changes may wish to skip directly to that discussion. Our objective here is broader and more academic: to link the history of institutional change in Table 13 to patterns observed by students of American political institutions. Putting Northwest electric power in context serves several purposes. First, it enables us to compare BPA's policies with those pursued by other government agencies. Second, the conflict that has surrounded these policies can then be seen as part of a wider reexamination of our faith in large-scale technology as a tool of social progress. Third, this political analysis establishes a conceptual base for evaluating proposals for reform in the next chapter.

In order to compare Bonneville's policies with those found elsewhere in government it is useful to draw upon the classification shown in Figure 17. This typology is adapted from the work of Theodore Lowi and Robert Salisbury.[2] The approach, developed originally by Lowi, is useful for our purposes because it pursues systematically the idea that political "expectations are determined by governmental...policies."[3] That is, that policies *already enacted* influence political behavior. This reverses the traditional perspective on policy as the product of political maneuvering, emphasizing instead the way in which existing policies shape political conflict.

This point of view led Lowi to an important insight: that the politics associated with different kinds of policies might reflect different divisions of the social order. "A political relationship is determined by the type of policy at stake."[4] Lowi laid out a three-fold division of public policies, later extended to four types. Each type of policy, he claimed, would be associated with a distinctive pattern of political cooperation and conflict. Policies for distributing public property or funds have been understandably popular among elected officials anxious to take credit for benefiting their voters. Distributive policies have been associated with "pork barrel" politics: broad legislative support accompanied by broad distribution of benefits. Redistributive policies are those

Figure 17. *Types of Public Policy (adapted from Lowi and Salisbury).*

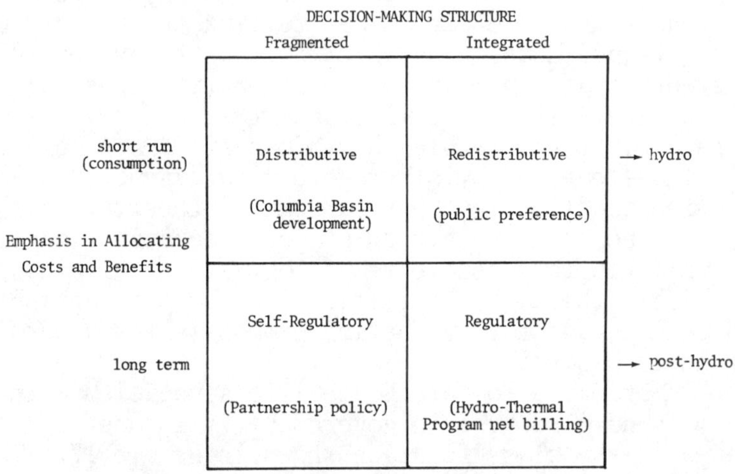

that transfer scarce resources, such as federal hydropower, from one group to another; such policies elicit sharp conflicts between winners and losers. Regulatory policies define specific governmental choices of who will be permitted to undertake regulated activities; competition among groups affected by regulation, while often intense, is normally not as divisive as redistributive politics. Finally, some groups obtain government sanction to operate on their own; self-regulatory policies are thus associated with political quiescence, since conflict in self-regulated arenas has been removed from the public sphere.[5]

This analysis is particularly useful in diagnosing conflict in circumstances in which more than one type of policy is involved. In such cases Lowi predicts that groups participating in the political struggle will have different expectations about the stakes of the game, its rules, and its satisfactory outcomes. Negotiations among interest groups are thus complicated by the need to reconcile differing conceptions of what is being negotiated and how compromise should proceed. The development of regional power legislation seems to have been particularly difficult in this regard. For our purposes, the strength of Lowi's perspective is that it allows an outside analyst to make informed guesses about how issues look to different participants in political debate. These in turn can be associated with differing patterns of coalition and styles of compromise.[6]

In Figure 17 we have identified four policies central to Northwest electric power with the Lowi-Salisbury types, and grouped them along two dimensions.[7] On one dimension we have sorted policies according to the way they allocate benefits and costs; one kind of policy focuses on the short run, while the other operates in the long term. On the other dimension we made a distinction based on the structure of decision making: some policies can be parceled out among relatively autonomous subunits, while others demand integrated decision making in the implementation of policy. These simple distinctions provide useful guides to the political dynamics underlying the chronology in Table 13.

BONNEVILLE AND THE IRON TRIANGLE

As we saw in Chapter II, The Bonneville Project Act proved to be a workable policy for the development of the Northwest hydroelectric system despite controversy over the legitimacy of public power and concern over the role of the federal government in financing hydropower. Both backers and antagonists of public power agreed on the immediate importance of low-cost electricity as a stimulus for economic growth. As we suggest in Figure 17, conflict in the hydroelectric period centered on how and whether national resources would be used to advance regional interests--that is, whether decision making would accord greater weight to decentralized concerns, or to central direction. Decentralized, regional priorities fared better. Both the activist New Deal stance of the early BPA and the quieter agency of the "partnership" period could live within the charter of the marketing agency defined in the Bonneville Project Act.

"Partnership" marked a federal withdrawal from utility affairs. This regional theme continued even after the resumption of federal activity in the 1960s in the Hanford New Production Reactor, the Columbia

River Treaty, the Pacific Intertie, and development of net billing in the Hydro-Thermal Power Program. In sum, the Bonneville Project Act endured for at least thirty years, until the Hydro-Thermal Program was initiated. By a more permissive accounting, one could argue that the original legislative framework seemed clearly appropriate until 1975, thirty-eight years after passage, when Phase II of the Hydro-Thermal Program was stalled by the lawsuit described in Chapter IV.

This stability of policy for a generation illustrates two trends. One, the resistance of entrenched public policy to control by elected officials is a theme much celebrated in American political studies. In addition, the durability of the Bonneville Project Act reflected the extraordinary consensus behind economic development as a social goal. Indeed, economic growth has been a force of global magnitude, including such profound changes as the European Economic Community, a point explored in a note at the end of this chapter.

The critical institutional lesson to be derived from the longevity of BPA's guiding policy is the influence and tenacity of government programs which have permanent, politically active clienteles and congressional sponsors. Among the institutional designs which have emerged in American government, none is stronger than the "iron triangle" of government bureau, client group, and congressional committee.[8] Bonneville, the utilities and direct-service industrial customers, and the Senate Interior Committee, chaired by Senator Henry Jackson, have formed a classic iron triangle. Their effectiveness can be seen in the decision to proceed with the New Production Reactor at Hanford, recounted in Chapter III. National policy--foreign affairs and national security in this case--served as an opening wedge in an argument to resume, in 1962, a regional program of federal investment in the Northwest.

Three hallmarks of the iron triangle suggest that the alliance is by no means moribund today. First, "personnel appointments often reflect private and congressional interests." [9] The current administrator of BPA is Sterling Munro, for twenty-two years an aide to Senator Jackson. Second, the iron triangle skews the incentives of political appointees toward the interests of agencies and clients, and away from the president's policies, leading "to White House complaints about... 'going native' in the bureaucracy....even the most presidentially minded political executive will discover that his own agency provides the one relatively secure reference point amid all the other uncertainties of Washington. In their own agencies, appointees usually have at least some knowledge of each other and a common identity with particular programs."[10] At BPA there is fierce pride taken in the regional character of the agency, and its independence from national dominance. "It's the only federal agency with a field office in Washington," in the words of one congressional aide.[11] And even former administrator Don Hodel remembered that, as a federal employee, he "felt that if the region could get its act together and get an adequate power supply lined up on a non-combative basis, we would then be in a position to say when they were trying to pass national grid legislation--hey, exclude us."[12]

Most of all, the iron triangle is sustained by interlocking material interests. When the Bonneville Project Act was passed in 1937, the policy of public preference tied the emerging public power agencies to Bonneville. In the press for aluminum production during the Second World War, the DSIs entered the region, and their large, technically desirable loads became an integral part of the BPA mission. Beginning in the 1950s, when the finite character of the federal hydro resource first became a serious concern, the investor-owned utilities serving the western part of the region joined the alliance. Indeed, BPA's ties to interest groups did not have to be cooperative in order to be politically useful. The traditional rivalry of public and private utilities, for example, has made their joint support of Senator Jackson's regional power bill that much more compelling to members of the regional congressional delegation.

The material importance of the iron triangle to its members reinforces their natural resistance to changes that are not of their own making. Bonneville has actively sought to influence legislation--not a civics role for a public agency. Partisans in the regional power debate commonly encounter BPA staff in the halls of Congress, and calls to high Bonneville officials at the Portland headquarters of the agency are answered by secretaries announcing that their bosses are in Washington, D.C. The close associations between BPA staff and staff members of the Senate Energy Committee (successor to the Senate Interior Committee) are unusual mainly in their openness. Most Washington-based federal agencies heed the formality that the President's Office of Management and Budget acts as a legislative clearinghouse for all proposals going to Capitol Hill from the Administration. In iron triangles information flows unhindered, but normally more discreetly than is the case with BPA.

The durability of the iron triangle linking BPA, the utilities, and Congress suggests that legislative revision of the Bonneville Project Act must satisfy the traditional alliance. A new regional power act would be a renegotiated treaty defining the relations among the utilities, DSIs, and BPA, ratifying traditions such as public preference and the principle of managerial independence among utilities, while permitting regional development to push on into thermal and other investment programs.

INVESTMENT WITHOUT TEARS?

From the standpoint of the utilities the need for a new charter has been building for twenty years. Electrical load grew steadily rather than abruptly, and the shortage of hydro which has become of immediate concern has been foreshadowed by problems in investment. (See the lower half of Figure 17.) The vital scarcity is that of thermal generating resources adequate to meet projected demands. This shortage has loomed over the Northwest intermittently since the partnership policy of the 1950s threw the region back on its own resources. The regionally based programs for power supply turned first to the mid-Columbia dams built by public utility districts in the 1950s, and then, as Democrats returned to the White House, invoked federal support again, in the Hanford Generating Project, and later the Hydro-Thermal Power Program and net billing.

The economics of thermal technology seemed, in the 1960s, to fit together naturally with the region-scale generating system of the existing hydro development. Yet, given the IOUs' traditional reluctance to permit a federal role in thermal power, federal financing of new generation could not take a direct approach. Federal participation in thermal plants--now confined to the Tennessee Valley--would require either special authorization for the Pacific Northwest, as is being sought now, or else an overturning of nationwide policy. So large a change would be, by common agreement, incompatible with the iron triangle's low profile and regional power base. The Hydro-Thermal Program, therefore resorted to a policy that provided federal economic assistance, using an essentially regulatory institutional structure. The commitment of federal acquisition through net billing pulled BPA toward a central management role. BPA purchase would pull it further down this path.

Although Bonneville purchase authority has been attacked as a "subsidy" to the Northwest, the politics of regional power suggest that proponents of purchase do not regard it as a distributive policy. The important economic fact is that rapidly rising thermal plant construction costs make it very difficult to quantify the actual benefits implied by purchase. Although interest rate differences can be estimated, purchase authority, by facilitating thermal projects, may induce premature construction or the building of plants in such haste that they do not perform as planned. The costs of these outcomes is quite large, although quantitative estimation is difficult.

What counts politically is that purchase authority is a mechanism for centralizing an economic decision, using government organization as the central agent. (Without the strictures of the Holding Company Act and antitrust policy, IOUs or other private sector factors might perform the functions of building and selling region-scale thermal facilities.) The utilities--fearing that BPA could undercut their managerial autonomy--have sought to limit the powers that BPA could exercise through purchase authority. Citizen group critics--fearing that centralized decision making will facilitate a bias toward thermal plant development, have sought to inhibit centralization itself. Thus the arguments over purchase emphasize the *regulatory* elements of its governmental functions.

The utilities' desire for new legislation to organize investment was not the only force pressing for change. For the iron triangle could no longer contain the issues to be addressed. The utilities themselves did not anticipate the broad public awareness of regional power issues that crystallized around the PNUCC bill. More fundamentally, the loss of a mass political consensus about the desirability of growth classed the utilities as a special interest, seeking ends which were no longer unquestioned, especially once political controversy broke out. Indeed, the linkage of electric power to non-economic issues such as conservation and environmental quality threatened to alter the very basis of the ties which have traditionally supported the iron triangle.

Politically, the most important change has been the full development of hydropower under conditions which have left hydro much cheaper than alternative supplies of energy. Under the preference policy ratepayers of publicly owned utilities will acquire property rights to all federally produced hydro. By the mid-'70s, the IOUs had essentially ceased to draw on federal hydro, and this contributed to rate disparities that dramatized the utilities' call for legislation. Once the economic effect of limited hydro became apparent, the divisions characteristic of redistributive politics were kindled once again.[13]

Uncomfortably for the publicly owned utilities, they were the "haves" of this situation, and the social justice which they had stood for in the New Deal now seemed to tell against them, in the cries of unfairness coming from representatives of investor-owned utilities and their consumers. In response the publics appealed to the historical mission of public preference--the program of federally sponsored development which had benefited the whole region. This reply, however, seemed to many to go past the issue of the moment--how to restore a condition of fairness which the shortage of hydro seemed to have upset.

A more acceptable response came in the Jackson bill, which was aimed at finding additional low-cost resources sufficient to serve the residential customers of IOUs. Thus the direct-service industrial customers of Bonneville emerged as the holder of a critical political counter: a pool of hydropower, commited on long-term contract, but available to resolve the complaints of IOU customers. Fortunately for the negotiations, moreover, the DSIs wanted something as well: long-term assurance of power supply, even at substantially higher prices. Thus, the way seemed open to resolve the rate disparity conflict while staying within the bounds of existing institutional arrangements.

THE LIMITS OF CONSENSUS

But as Chapters IV and V indicate, resolving economic complaints may no longer be enough. Seen from this political and institutional level, the controversy in the Pacific Northwest is one of a set of disputes over how large-scale technologies with diverse economic, social, and environmental consequences should be brought under some form of governmental control. In part, these concerns are as old as industrialization itself,[14] but the extensive reliance upon organizational design as a solution, together with governmentally imposed regulations and transfers of economic resources, have come into increasing prominence since the Second World War.

The postwar optimism that technology in all its manifestations was a boon to mankind gave way during the 1960s to a growing realization that technological development could itself constitute a broad social problem. Two of the most ambitious analyses, contained in the works of John Kenneth Galbraith and Daniel Bell, outlined a subtle but pervasive contradiction in advanced industrial societies.[15] This is the conflict between technological rationality and a simultaneous cultural emphasis on self-expression and individual fulfillment.

The technological rationality of the Hydro-Thermal Power Program was analyzed in Chapter III, drawing upon Galbraith's perception that large technologies require large organizations--organizations which acquire a life and purpose of their own. These purposes are under increasingly sharp attack from self-styled public interest advocates, who call for a more humane and human-scale order of life, and who suggest that it be pursued through the mechanisms of a revived citizen-based politics. This is the neo-Progressive orientation of Seattle's Energy 1990 decision. From the standpoint of the utility establishment, Energy 1990 was a romantic, self-indulgent, and ultimately irresponsible disruption of an orderly investment procedure.[16]

The appeal to public values of those who question monolithic technology is immediately attractive to the denizens of a modern society: we are all laymen confronted by Juggernaut in most of our lives. This intuitively persuasive--hence broadly shared--mistrust of large-scale technocratic institutions is the basic political resource drawn upon by public-interest advocates. The bias toward mistrust becomes a tendency toward deadlock, deadlock that can permit the wider articulation of issues basic to neo-Progressive participation. Moreover, deadlock is itself a bargaining lever, even though deadlock can be irreversible in the short run. It has often been true, in any case, that public-interest advocates have concentrated on raising questions rather than suggesting alternatives, disdaining the considerations of economic cost, engineering feasibility, and broad political acceptance that form the pragmatic canons of the technocratic elite. Edward Banfield, a defender of that elite, concludes a review of municipal reform with the sour assessment that

> the range of feasible measures for dealing with the serious problems of the cities is much narrower than one might think, and, second,...within this range hardly any of the feasible measures are acceptable. If what is feasible is not, in general, acceptable, the reverse is also true: what is acceptable is not, in general, feasible. Moreover, government seems to have a perverse tendency to adopt measures which...are the very opposite of those that one would recommend. The reasons for this perversity are to be found in the nature of American political institutions and, especially, in the influence on public opinion of the upper-class cultural ideal of "service" and "responsibility to the community."[17]

Reformist politics has significantly affected the political deadlock which stalls the development of large-scale technologies. Rising electric power rates in the Northwest worry politicians. The broad impact of these increases could have electoral repercussions if given voice by skilled opponents. Opposition is often articulated through a reformist critique, even though the reforms advocated by proponents of citizen participation are essentially ideological rather than economic. Such ideological claims cannot be readily compromised, especially when the compromises are discussed amid complex, protracted negotiations which may come apart later anyway.

> The upper-class ideal recommends participation as intrinsically good, but unfortunately, the more participants there are, the larger the number of issues that must be dealt with and the greater the disagreements about each. The ideal also requires that issues be settled on their merits, not by log rolling, and that their merits be conceived of in terms of general moral principles that may not, under any circumstances, be compromised.[18]

Nonetheless, the raising of moral claims without an effective reply from the technocratic establishment has left politicians convinced that some form of citizen representation is necessary.

The problematic character of large-scale technology has three political implications which have become prominent in the Pacific Northwest. First, the apparatus for making legitimate political decisions has weakened. Second, and simultaneously, the administrative burden on government has increased. And third, issues which were formerly regarded as independent have been deliberately linked together in the public decision-making arena.

The weakening of the legitimate structure of representation is to some extent a spin-off of the moralistic style of politics which created the broadened agenda. Consensus has been hard to regain, especially when national legislative leaders are far removed from the scene of conflict. Indeed, as we pointed out in Chapter V, the congressional hearings on the PNUCC bill provided the first occasion for many members of the Northwest congressional delegation to understand the political forces at work in the regional power debate. Worse, national representatives 3,000 miles away from their constituencies must inevitably rely upon interest-group leaders, leaders whose own capability to represent their followers has become less certain. Thus, it may be impossible in the short run to channel all political competition through interest groups: the leaders of both the environmental movement and the utilities find themselves, for different reasons, unable to speak authoritatively for their constituencies even if they try earnestly to represent them.

The conflict over electric power has also revealed an institutional weakness in the governmental structure. The disjointed apparatus of federalism, a helter-skelter skein of intergovernmental cooperation among state and local government and federal agencies, has evolved as a mechanism for transferring federal resources to lower levels of government. It should be no surprise that the structures of intergovernmental cooperation set up to deliver distributive policies turn out to be poor at regulatory administration.[19] Moreover, bureaucrats have been taught to think of themselves as neutral, objective public servants. They are bound to fail when they must represent newly controversial agencies under circumstances where different levels of government may all be responsive to the same mass constituency, but different interest groups.[20]

The result of these difficulties is a weakening of the legitimacy accorded official decision makers. Perhaps worse, administrative organizations on the scene are drawn into political roles as they seek to be responsive, but have no clear guidance about whom to be responsive to.[21] Former BPA administrator Don Hodel, in his speech attacking environmentalists as "prophets of shortage," demanded help from the "organizations and individuals who have taken the lead in anti-energy litigation....since they and their no-growth ideology are partly responsible for the shortages we confront, they should help make the hard decisions in sharing what energy is available."[22]

A second political accompaniment of large-scale technology is the steady expansion of the administrative burden on government. Technological complexity and economic growth widen the possibilities for conflict in civil society. Government is accordingly called upon to act as referee with increasing frequency. Regulatory functions grow and, as they expand in size, spawn bureaucracies:

> Even as late as the middle 1950s the federal government had a major regulatory responsibility in only four areas: antitrust, financial institutions, transportation, and communications. In 1976, eighty-three federal agencies were engaged in regulating some aspect of private activity. Thirty-four of those had been created since 1960 and all but eighteen since 1930.[23]

In short, political determinations of functions to be served by the government lead rapidly to bureaucratic organization.[24] Energy conservation, an objective which combines regulatory policies with government subsidies, seems to be the next growth area in government presence.[25]

The third political consequence of large-scale technical development is issue linkage. Independence of issues is crucial to the stability of the iron triangle alliance; the scope of the controversy must be narrow enough to settle disputes "in the family." Yet large-scale technological development ties issues together. In the Pacific Northwest, state governments were until recently content to leave well enough alone in electric power, although public enthusiasm for environmental quality and concern over land use had gradually led to governmental encroachment on utility freedoms. But the dramatic economic shifts accompanying thermal power development have brought the state governments on stage permanently. Thus formerly independent authorities come into conflict.

As issues have become linked to one another, the goals of regional power development have expanded with the number of participants and become inconsistent: although conservation and greater state government involvement in utility affairs are now sought by the utility establishment, both these objectives must be reconciled with the continuing commitments to corporate growth and managerial autonomy. The result is turbulence: "The number of actors is very large; each actor pursues a variety of objectives which are mutually incompatible, but each is unsure of the trade-offs between the objectives; each actor is tied into a network of interdependencies with other actors who are as confused as he."[26]

RESPONSES AND IMPLICATIONS

Despite the turbulence which has afflicted the legislative process, one can capture much of what is at stake in the regional power debate in two somewhat oversimplified positions:[27]

> *Model I (Technocratic Elitism):* There are overriding common interests in regional power planning, interests that primarily affect a limited number of well-organized groups, notably utilities. The leaders of these groups can act knowledgeably on behalf of their constituencies. As a result, negotiation can lead to agreements acceptable to all participants.

> *Model II (Unstable Pluralism):* The scope and depth of conflict in regional power planning is significant and growing, involving many viewpoints beyond those of the utilities. Those who have participated in negotiations have not always understood or represented the interests of their constituents. As a result, public disagreement is common and those who call for consensus either do not understand the situation, or are trying to shut off debate.

Table 14 provides additional details fleshing out these two modal positions. Note, in particular, that the Technocratic Elitist position emphasizes *administrative* solutions guided by allocations of *economic* costs and benefits. By contrast, the Unstable Pluralist view takes a more *political* stance about the function of new decision-making structures--that they legitimate changes in who makes and influences regional power planning decisions; the emphasis here is on developing *public purposes* to be served by regional power development.

The evolution of recent legislation follows the model of Unstable Pluralism. Indeed, the odyssey of the regional power bill is an excellent illustration of William Gamson's remark that "efforts toward broader planning...may have the incidental and unintended consequence of increasing the degree of competition."[28]

Table 14. *Two Models of the Political Process in Regional Power Legislation.*

	Model I. Technocratic Elitism	Model II. Unstable Pluralism
Central Problem	Power supply	Regional future
Legitimate Participants	Utilities; Congress, DSIs	Also governors, local government, citizen groups
Key Issues	Financing of new generation allocation of federal hydro	Conservation; allocation; forecasting; governance
Mode of Resolving Conflict	Allocation of *economic costs* and benefits within a clearly defined *administrative* structure	Changing participants, decision-making structure, and methods of selecting leaders, leading toward a new political consensus on uses of energy and other natural resources
Predictions and Prescriptions	Reliance on administrative solutions Basic driving forces are economic Technological necessity and complexity determine outcomes High risk of chaos if no consensus	Political solution needed Public purposes need to be redefined as guidelines Low risk of disaster or even shortages

Yet, Model II does not provide a modus operandi for those pursuing regional solutions to the problems of the utility industry. Those seeking a regional bill have suffered the frustrating fate of operating according to a model of political process unsuited to the political environment.[29] Is there a choice between seeking to *impose* Model I and being paralyzed by Model II? This is a key question in devising a new institutional structure which can be politically feasible.

Legislative Approaches

The PNUCC and Jackson bills might have threaded their way between these unpalatable alternatives if the legislation had been proposed in early 1977, as the PNUCC drafting group had originally hoped. But tactical errors critically damaged the process by which ideas were introduced, widening disputes within the technocratic elite and permitting the entry of other voices.

The key weakness of the PNUCC bill, as illustrated in Figure 18, was the attempt to frame a legislative package that would preserve a fragmented decision-making structure. This would have protected a regionally oriented BPA and insured utility autonomy in this policy arena, in the tradition of Technocratic Elitist politics. Yet the PNUCC bill did not address the impact of rate

Figure 18. *Issues and Legislative Responses.*

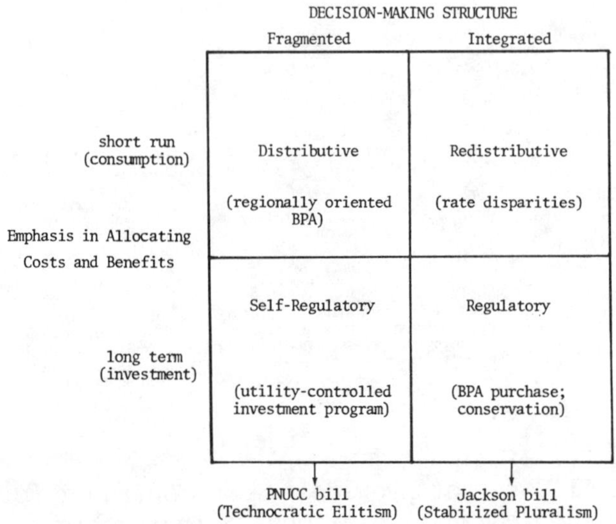

disparities in a form which was politically comprehensible. As the account in Chapter V showed, the rate pool structure of the PNUCC bill raised questions about public preference among public utilities, while failing to stem broadening concern about rate disparities separating public and private utility customers. Thus, even though leaders of the various factions had agreed on the language in the PNUCC structure, some of their followers were reluctant to let the matter rest there, particularly because the package had the aura of a special-interest bill. As with environmental opposition to the Hydro-Thermal Program, delay fostered increasing doubt, and the PNUCC bill became enmeshed in a widening circle of complaints, complaints fed in many respects by the elaborateness of the structure proposed.

A different strategy was adopted when politicians drafted legislation--an attempt to stabilize the Unstable Pluralism then evolving. The Jackson bill sought to integrate policy concerns from a number of different constituencies and deliberately widened the sphere of decision making to encompass non-utility actors. The DSIs' willingness to transfer the economic benefits of federal hydro to the residential customers of IOUs tackled the rate disparity problem head-on. At the same time, by tying BPA purchase authority to strong backing of conservation and renewable resources, the Jackson bill sought to accommodate those urging changes in the utility-dominated decision process. This strategy was designed to build a larger coalition within the region.

Regional consensus was explicitly intended to be a prelude to enactment of a Northwest power bill in Congress. Jackson would pass legislation favoring his home region, in exchange for his support to other legislators on other issues. It was crucial to this strategy that wider national concerns not be raised, in particular the sanctity of the preference clause and the appropriateness of BPA participation in financing regional power plants. Yet both these issues had surfaced already in the controversy over the PNUCC bill. Thus even the integrated decision structure, which might once have sufficed to build regional consensus, could not serve to generate a national coalition. This was especially evident in the House of Representatives in 1978, where there was no Northwestern legislator with sufficient influence to make logrolling deals of the kind Henry Jackson was prepared to make in the Senate.

By mid-1979 when this book was completed, the political configuration remained much the same. Although the Jackson bill passed in the Senate, the fate of a regional power bill in the House of Representatives remained uncertain.

Implications for Design

The mutual suspicion which pervades the language of S. 885 suggests that even if a regional power bill is passed it will not enjoy the longevity of the Bonneville Project Act. Legislation, in short, will not solve the problems which led to the regional power debate in the first place. We should accordingly recast our analyses of institutional process with an eye toward strengthening the ability of existing organizations to work within a new legislative charter. Two major developments must be considered.

(1) With the rise of controversy, the context of electric power planning has shifted crucially:

> That context is exceeding political, imbued with clashing interests....Hence there is every reason to treat our administered organizations as polities that respond to external demands, and that may develop into independent centers of power if the "proper" interaction with those demands is achieved by administering the organization in a certain way. Administration, then, ceases to be an internally directed, management-dominated concern, and becomes a politically adaptive pursuit in which leadership is crucial.[30]

Thus, more than a legislative renewal of mandate is needed: *there must be institutionalization* as well. *Institutionalization*, as we use the term, is the process of clothing organizational entities--either new or extant--in political values and social meanings. The iron triangle of the technocratic elite has been the institutionalized form of the Bonneville Project Act. The challenge now is to adapt to a different technological and social setting. The institutionalization process requires politically coherent organizational purposes, an elite group from which the leadership of that organization may be recruited, and the support--including opposition--of interest groups with popular bases.[31]

Institutionalization is closely bound up with leadership, as we shall discuss in the next chapter. One of the positive aspects of the deadlock that has developed in the region is that it has provided a limited but genuine opportunity for new entrants to the regional debate to be tested for leadership potential. Although the probability is not yet comfortably high, the possibility of recruiting intelligent leadership able to make use of a revised legislative framework for regional power planning seems brighter by far than it was several years ago.

(2) The *"shadow of the bureaucrat" looms over all possible alternatives.*[32] The tenacity of the Technocratic Elitist view is attributable in large measure to the pressures and situations faced by career officials

in large organizations--utilities, BPA, and even congressional staff members. These mid-level persons face increasing demands for action, but are cut off from political accountability by virtue of their organizational position, and the complexity of issues under debate. It is a battle, as Haas's portrayal of turbulence suggests, in which there is no effective reconnaissance of the battlefield. This situation is likely to continue, even with a new legislative framework. The implementation of new policies, accordingly, will be constrained by the difficulties of operating in a multi-issue arena of moderate--and occasionally high--controversy.

(3) It is therefore clear how institutional design should attempt to synthesize a middle path between the models of Technocratic Elitism and Unstable Pluralism: there must be a combination of *administrative design, together with a political process which can articulate public purposes,* and hold officials politically accountable for striving to meet them. The Jackson bill remains ambiguous on the choice between Technocratic Elitism and such a Stabilized Pluralism. It seems unlikely, in short, that any regional power bill could authoritatively resolve even the most important conflicts for very long.

Although the dynamics of the political situation have led implacably toward deadlock, we have also found reason to believe that limited institutional change can shift the momentum into more constructive paths. To do so will require, however, the harnessing of significant political energies, together with a mastery of the complex configuration of interests at stake in the regional debate. It is to a consideration of how such power can be marshalled, and for what ends it should be used, that we turn next.

NOTE: REGIONAL INTEGRATION--AN ANALOGY

The development of the regional electric power supply in the Northwest parallels in an interesting way the phenomenon called "regional integration" in the world economic and political system. Twenty years ago, as Western Europe recovered from the ravages of the Second World War, the development of advanced industrial societies in European countries led to tensions similar to those arising in the Pacific Northwest. The scale of technological and economic enterprise demanded cooperation on a scale much larger than the European nation-state. The momentum of economic development made it politically sensible for nation-states to cede part of their sovereignty to multinational governmental institutions like the Common Market.[33] Yet as the transcending of national sovereignty proceeded, a surprising pattern emerged: some international organizations flourished, but national governments did not weaken. Although national sovereignty was dysfunctional for some purposes, national identity proved to be durable, and more important, national governmental leaders retained the ability to surrender their independence in limited and usually reversible ways.

To see what this means for the Northwest, consider the following analogy. In the regional economic and technical structure of regional electric power development, the role of nation-states is played by the utilities and BPA. Even though the utilities derive their formal authority from the state and federal governments, the political roles and policy positions of the utilities do not necessarily reflect their dependent status, especially when utility and governmental goals are aligned by forces such as the iron triangle.[34]

"International organizations" like the Pacific Northwest Utilities Conference Committee share the tasks of inter-utility coordination with state governments which act, in this context, somewhat like international organizations. (Indeed, state utility regulatory commissions have usually been politically weak and isolated, rather as international bodies have been.) This institutional analogy permits us to adapt the literature of international integration and to suggest some implications for the Pacific Northwest.

The first set of implications stems from the surprisingly close parallel between the dominant role of BPA in the hydroelectric period and the dominance of the United States in the post-World War II global system.

The early history of Bonneville was shaped by the availability of vast quantities of low-cost federal hydro power, a resource which was bent to the purpose of building a "Free World" of community-oriented public agencies. This aggressive policy, paralleling the containment strategy of the Cold War, has left a lasting imprint on the investor-owned utilities, analogous to the historic fear of encirclement in the Soviet Union.

The overriding influence of BPA was accompanied by the growth of organizations well-suited to take advantage of its supremacy: the direct-service industrial customers, drawn to the Pacific Northwest by the availability of assured power supplies at low cost. By making exceptionally full use of the hydro resource, the DSIs supported by their very presence the wisdom of large-scale federal investment. This symbiosis between DSIs and federal government broadly resembles the mutual support of the national government and the business firms that have recently come to prominence as multinational corporations.[35]

By the 1960s, the technical advantages of the DSI electrical load became increasingly important as a means of "firming up" hydropower--and thus extracting additional power from the fluctuating Columbia River. Indeed, BPA has found it technically efficient to provide power to the DSIs, according these customers higher priority than IOUs for firm power. This action, in a context of widespread public misunderstanding of the technical and economic role of the DSIs, has led recently to an attack on the DSI contracts as a bastion of unwarranted privilege. The public misgivings about multinational corporations is strikingly similar: note especially the concern voiced by organized labor that international corporations are "exporting" jobs by moving manufacturing to nations with lower wage scales. The high energy consumption and low labor use of the DSIs has been cited disapprovingly--although no one has suggested a labor-intensive means of making aluminum.

The analogy with international relations leads us to see the partnership policy of the Eisenhower administration as the abdication by BPA of "strategic superiority," as local utilities developed the principal remaining Columbia River hydro sites and non-federal hydro became available to investor-owned utilities. In an era of rising generation costs, non-federal hydro has become an economic shelter for investor-owned utilities such as Washington Water Power and Puget Sound Power and Light.

In the international arena, as the Soviet Union acquired strategic parity with the United States in the 1960s, it became both possible and necessary to relax tensions, both in the military sphere, in the strategic arms limitations talks, and in the political sphere, in the wavering pursuit of *detente*. The necessity for rapprochement between public and private power--sensed as early as 1962 by Senator Jackson in the case of the Hanford N-reactor--was realized in the Hydro-Thermal Power Program of 1968.

The full development of Columbia River hydro resource implied, in political terms, the erosion of BPA's superiority in the region. Bonneville would--and will--continue to play a central role in transmission and in management of the Federal Columbia River Power System, including the upstream development established by the Canadian Treaty. But centrality no longer translates into influence, because Bonneville can no longer utilize its strength. Bonneville's dilemma parallels, in some ways, the frustrations of the United States in the Vietnam War. With full development of the hydro potential--and lacking authority to take the lead in thermal development--Bonneville's relationship with its customers is one of routinized interdependence. While in theory BPA could selectively cut off individual customers, in practice, the prevailing standards of fairness--both administrative and political--make cutoff unlikely; even the threat of cutoff is not credible. The amount of power potentially available to BPA is diminished by the difficulties of its exercise.

The dwindling power of BPA is most evident in its inability to assist in the financing of thermal power stations. Without a mechanism like net billing to pool regional resources for expanded power supply, each utility is put under pressure to fend for itself. As Gilpin's analysis of the decline of U.S. dominance suggests, the global economy envisioned by advocates of liberal trade arrangements is giving way to mercantilism: the assertion of national interest, even when economic gains could be won by multi-national cooperation.[36] The issuing of Notices of Insufficiency to its preference customers in 1976 was BPA's Vietnam: the visible delineation of the limits to regional leadership. The scramble for utility self-interest was on.

From this perspective, the conferring of authority to purchase power and to regulate conservation is a Bonneville "rearmament" scenario: purchases from plants still to be constructed, as well as supervision of regional conservation, involve many discretionary decisions, discretion available for use by the Bonneville administrator. Thus, thermal power and regional conservation could become important political resources in the appropriate institutional setting. But, as with the dubious promise of expanded U.S. defense budgets, it is far from clear that rearmament by itself can succeed in rebuilding institutional strength. It may be, nonetheless, an indispensable part of a redefinition of the agency's regional mission.

This historical parallel is unfortunately of little predictive value, since the difficult positions of the United States in the world order and of BPA in the Northwest stand at roughly the same impasse; neither is a "leading" indicator. But a closer examination of the internal structure of regional integration theory suggests other linkages that are helpful in understanding the Northwest power situation.

As Ernst Haas has recently pointed out, the interdependence of national societies is genuine and growing, but that does not mean that national sovereignty must increasingly be superseded. The creation of institutional structures at the international level may not be necessary to form the coordinating structure needed to harmonize the interests of nations, their citizens, and component business corporations.[37] This finding reminds us of the utilities' determination to retain their managerial independence in the face of region-wide interconnection; how this might be turned to advantage is accordingly an important issue for institutional reform.

In an analysis of changes in the international order, Robert Keohane and Joseph Nye have made a number of generalizations of the ways in which international cooperation changes political relations.[38] A number of their observations parallel the broadening of the agenda in the Pacific Northwest. The goals pursued by utilities and other players in the regional power game vary by issue area, and are frequently mutually inconsistent. The demands of regional coordination, in particular, often conflict with the objective of utility independence. Contacts between leaders fade in importance as the political situation becomes more fluid and confusing; correspondingly, contacts at the "working level" among middle-management officials become more influential. Inter-utility coordinating organizations, including state government agencies, become an effective arena for setting the agenda of debate, and a mechanism for coalition-building by weaker players.

The analogy between utility politics in the Northwest and international politics is imperfect. The most important shortcoming is that the federal government constitutes a supreme authority--the analogue of a world government, able to impose a solution on all the utilities and other organizations participating in the debate. As Chapter V makes clear, however, national government has not been willing to impose a solution on the Northwest. Indeed, the amendments to the Jackson bill proposed by the governors of the Northwest states provide for greater regional autonomy in the future--an autonomy that can presumably last beyond the tenure of Senator Jackson.

CHAPTER VII
Institutional Alternatives

> Start from the siting of a power plant, the internal combustion engine, or the double helix, and one comes in time upon the interrelated ordinances of nature, man and God.[1]
>
> --Elting Morison

The hydropower that has for four decades been the mainstay of electrical supply in the Pacific Northwest is now largely developed. Further growth in demand will have to be met from coal, nuclear, conservation, and renewable and unconventional sources. All the supply alternatives represent costly commitments of social and natural resources. Since large-scale federal hydro development began in the Northwest during the Depression, the region has changed dramatically; federal sponsorship of Columbia Valley development has played a prominent role in this historical development. The shift now underway in the technology and economics of power supply presages changes of similar scope for the future.

The utilities' West Group Forecast projects additions to the regional electric power supply of just under 26,000 megawatts of generating capacity by 1998.[2] At today's prices--a conservative estimate of future costs--this would be a capital investment of over $30 billion. These additions--largely of thermal generation--would somewhat more than double the average power available in the region, but the increment is expensive. For comparison, Bonneville anticipates that investment in the federal hydro system in 1998 will amount to only $8.5 billion.[3] Thermal power will dwarf the existing system in cost.

The impact on ratepayers will be substantial. In an analysis for the Jackson bill BPA projected that by 1994-95 the capital costs of generating facilities owned or financed by the federal government would increase by 500 percent. Wholesale power costs for publicly owned utilities, now averaging 3.5 mills/kwh, would triple in the same period from 1980-81 to 1994-95.[4] Since these rates are scheduled to double before 1980, wholesale rates will increase approximately sixfold between 1979 and 1995.

As these figures make plain, choices made in the next several years will commit the Northwest for at least a generation to major new sources of electricity and their environmental, economic, and social consequences. Those commitments will not solve all the problems of electric energy--any more than the creation of the Bonneville Power Administration settled all the issues in 1937. But they will shape the possibilities decisively--much as BPA has done.

In this context, the need for regional management of energy and resources appears self-evident, although how responsibilities should be divided is far from clear. BPA, the existing regional energy agency, has performed an indispensable mission. But the growth stimulated by federal investments in the Pacific Northwest has broadened the scope of public responsibility and made social choice complex and unwieldy.

Bonneville is likely to retain a major role, though its mission and functions will change under the pressure of circumstances and perhaps congressional legislation. Yet there are important responsibilities that BPA cannot or should not bear alone in planning and managing the multiple considerations of regional power development. In an era plagued by inflation and struggling with conservation, and in a region with great strengths in agriculture, forestry, recreation, and manufacturing, the governance of electric power should reflect shared leadership capable of balancing the complex demands of an interdependent society. This is the challenge of institutional design.

Institutional design is a political choice. In the Pacific Northwest, four factors shape the spectrum of available alternatives: the federal presence; the expanded scope of conflict; an associated widening of state and local government involvement; and the utilities' defensiveness.

The principal institutional fact is the federal presence. From the thirty hydro projects that have domesticated the Columbia River and its tributaries to the 116 public power agencies, federal architecture has shaped the Northwest power supply system and the informal social structure of Bonneville's relations with its clients and patrons. One of history's ironies is that the federal preference policy should stand as a barrier to regionally coordinated power development, for the preference clause in the Bonneville Project Act was intended as a spur to "encourage the widest possible use of all electric energy." Moreover, within two decades --unless the most outspoken environmentalists are right and everyone else wrong--growing demand, supplied by thermal power, will have diluted in a flood of rising cost the precious pool of low-cost federal hydro. Everyone's electricity prices will rise rapidly in the decades to come; that is the real cost of growing demand.

The preference clause and sharply higher thermal power costs are not the only complications of institutional design. The agenda of debate has broadened. This is most evident in the new prominence of energy conservation and environmental protection. Without the difficulties of the nation's energy policy, together with growing awareness of the fragility of the "livable" Northwest, these goals might be given short shrift--as they were when the Hydro-Thermal Program was designed a decade ago.

State and local governments have also emerged, if rather unevenly, as centers of political concern about regional power planning. Since the establishment of state energy offices in the mid-1970s, state energy bureaucracies have become new channels for public communication, separate from the controversies of power plant siting and the internecine complexities of public utility regulation. Oregon's Department of Energy, founded in early 1975, has been an important source of analysis and critique. And both Seattle and Portland have played visible if not decisive roles in congressional legislation.

Fourth, utilities, both public and investor-owned are fearful of being losers in the politics of power--losers to one another, and to new, non-utility voices demanding entree to the traditional "iron triangle" of utilities, BPA, and their congressional allies. Utility "capitalism" has produced a situation in which government-guaranteed purchase of electricity seems to be a precondition for private enterprise--including investors in WPPSS revenue bonds--to risk its capital.

These four forces urgently press for new congressional legislation, to reconstitute a workable framework for regional power planning. Such a framework would revise and supplant the Bonneville Project Act of 1937, and the legislative accretions it has acquired since then. Such a framework should provide for explicit oversight in those areas where broad, long-term interests of the Pacific Northwest and the nation are at stake.

But whether or not Congress manages to build such a framework, the four factors that shape regional power politics will persist. A revitalized congressional mandate will not eliminate conflict, though it may alter the coalitions and issues in contention. Such a management regime must accordingly possess flexibility sufficient to cope with the uncertainties that lie ahead. That flexibility, moreover, must be carefully nurtured in a political environment in which high uncertainty has become pervasive.

Our aim in this chapter is to appraise three principal institutional alternatives. The first two set conceptual benchmarks: continuing deadlock, of the sort we are currently experiencing; and the defederalization of the Columbia River Power System, leaving the Pacific Northwest states to squander or husband their federal inheritance.

Third, is the strong-BPA alternative envisioned in the Jackson bill--a Bonneville Power Administration equipped with the organizational tools and governmental authority to lead the utilities of the region, including a revised approach to the allocation of federal hydropower. As we argued in Chapter VI, the linkage of electric power development to a broad agenda of social concerns means that no single leader, not even Bonneville, can control regional power planning with the effectiveness of the New Deal BPA. We therefore believe it important to improve the ability of the agency to recognize and to manage in an accountable fashion the conflicts that at once obscure and constitute the region-wide interests of the Pacific Northwest.

These alternatives focus upon institutional arrangements for a region-wide power supply system, together with region-wide financing and coordination of energy conservation to manage growth in demand. This emphasis on governing large-scale technologies inevitably underemphasizes the role of decentralized, small-scale technologies like windpower or passive solar. Significant adoption of these alternative energy technologies would alter the outlook for energy conservation and, eventually, the scope of regional power planning itself.[5] Thus, while a region-scale power system will dominate electricity supply for the foreseeable future, its social and political significance could dwindle.

CONFLICT MANAGEMENT

As the evolution of the Jackson bill demonstrated, the scramble for political compromises pushed out elements of a comprehensive institutional design. Where the earlier PNUCC proposal included a regional organization and explicit language for power contracts in addition to congressional legislation, the Jackson bill was saddled with amendments drawn up by the Public Power Council which reaffirmed the preference policy. While it still contained language suggesting ways to manage in the future, the Jackson bill also became a record of the stakes which contending interests found it possible to have Congress ratify. An indispensable aspect of managing the regional power supply system in the future must include means to recognize and resolve conflicts.

We place emphasis on the recognition of conflict in the belief that the struggle to rewrite the Bonneville Project Act reflects tensions which are real and durable; they cannot be extinguished by Congressional fiat even if it were forthcoming. "The greatest source of tyranny," Don Price wrote, "is the conviction that there is a single way of determing truth."[6] Conflict is, accordingly, not necessarily a sign of trouble, except for those currently in charge. Figure 19 presents a simple model of conflict and its consequences.[7] Conflict arises naturally from the implementation of policies. Stable policy frameworks, such as the Bonneville Project Act during the hydroelectric period, can accommodate conflict through changes in policy, organization, or personnel. When conflict resolution fails--the situation called deadlock--there are two possibilities. First, deadlock may make conflict worse. Alternatively, one party to the conflict, normally the weaker, may have its views repressed or ignored. Repression can postpone conflict.

Figure 19. *A Simple Model of Conflict.*

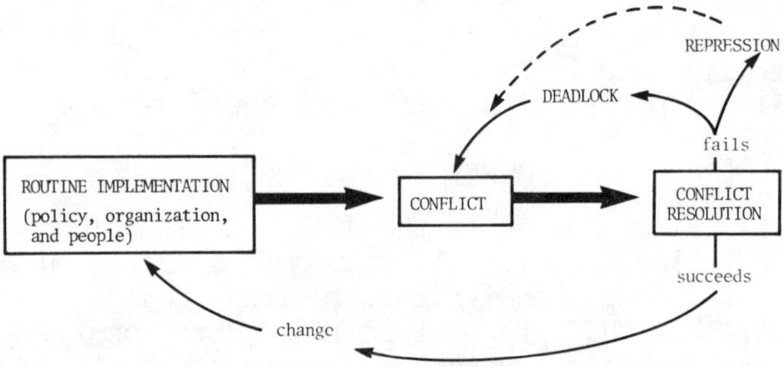

When there is conflict but decisions must still be made, the style of decision making must adjust to lack of consensus, as shown in Figure 20. When consensus is high, as in the day-to-day operation of the Federal Columbia River Power System, decisions are routine: the style is administrative--working out answers to problems using clear-cut formulae and rules. Bureaucratic organizations marshal information well in this setting. When agreement over the way things work is weak, judgment is needed. No one has a corner on wisdom, so some means of aggregating the judgments of colleagues is needed to arrive at a course of action in the face of uncertainty. The drafting of the PNUCC bill followed this pattern, though with substantial friction over who

Figure 20. *Decision Making With and Without Consensus.*

	AGREEMENT ON ENDS: high	AGREEMENT ON ENDS: low
AGREEMENT ON MEANS: high	ADMINISTRATION, via bureau structure (routine power supply)	COMPROMISE, via bargaining among representatives (regional delegation hearings on PNUCC bill)
AGREEMENT ON MEANS: low	COLLEGIAL JUDGMENT, by expert peers (PNUCC bill drafting)	deadlock, leading to WIDENING CONFLICT or INCREMENTAL "solutions"

Source: Adapted from Ernst B. Haas, *Beyond the Nation-State*, p. 105, citing James D. Thompson and Arthur Tuden, "Strategies, Structure and Processes of Organizational Decision," in *Comparative Studies in Administration*, ed. Thompson *et al.*

would be included among the experts to be consulted.

People can disagree over ends as well as means. Here, the problem is not uncertainty about the environment, but disagreement over purposes. The decision-making style must therefore harmonize competing goals through compromise. As the PNUCC bill became the focus of public conflict in Congress, members of the Senate and House sought to define contending ends and the compromises which could build coalitions sufficient to pass a regional power bill.

When disagreement over means is accompanied by disagreement over ends, there is stalemate. If a clear decision is to be made, a new vision of ends and means needs to be built. This is the purest, most challenging form of political leadership. But unclear decisions can often be made, especially ones that change existing relations only incrementally.[8] As Figure 19 suggests, an incremental response can be a form of repression.[9] In addition, the temptation to move in the direction indicated by the arrow in Figure 20--toward incrementalism--is a powerful one in American politics: to paper over disagreement over ends by seeking agreement over means. This was the political strategy of the Hydro-Thermal Program.

By linking together the model of Figure 19 to the typology of Figure 20, we may develop a general perspective on how conflict may be managed. This is shown in Figure 21.[10] The lesson drawn from Figure 20 is that different decision-making structures are needed to resolve different kinds of conflicts. Thus as we indicate on the right side of Figure 21, the implicit political models we called Technocratic Elitism and Stabilized Pluralism in Chapter VI invoke different kinds of responses to the "problem" of regional power planning. These different responses reflect, of course, different conceptions of what the problem is. It is interesting that legislative compromise is the style of decision making which both models agree is an element of conflict management. Their agreement on this point, combined with their continuing disagreement on much else, suggests again that legislation is not the whole answer.

Managing conflict under conditions of high technological complexity and political uncertainty is the task of regional leadership. It is a task for which none of the principal contenders has much relevant experience.

INSTITUTIONALIZATION

Under the conditions envisioned for the Hydro-Thermal Program a decade ago, regional challenges were to be faced by utility administration outside the turbulence of the political arena. The continued expansion of power supply was non-controversial. Regional power planning needed little governance beyond efficient administration--a view most recently reflected in the PNUCC proposal, with its emphasis on utility representation in forecasting and "gating." But as Ernst Haas remarks,

> the process by which a given activity becomes non-controversial is itself a political matter, derived not from initial consensus but from initial conflict....it follows that a purely rational decision-making model is...inappropriate.[11]

Figure 21. *A Model of Conflict Management.*

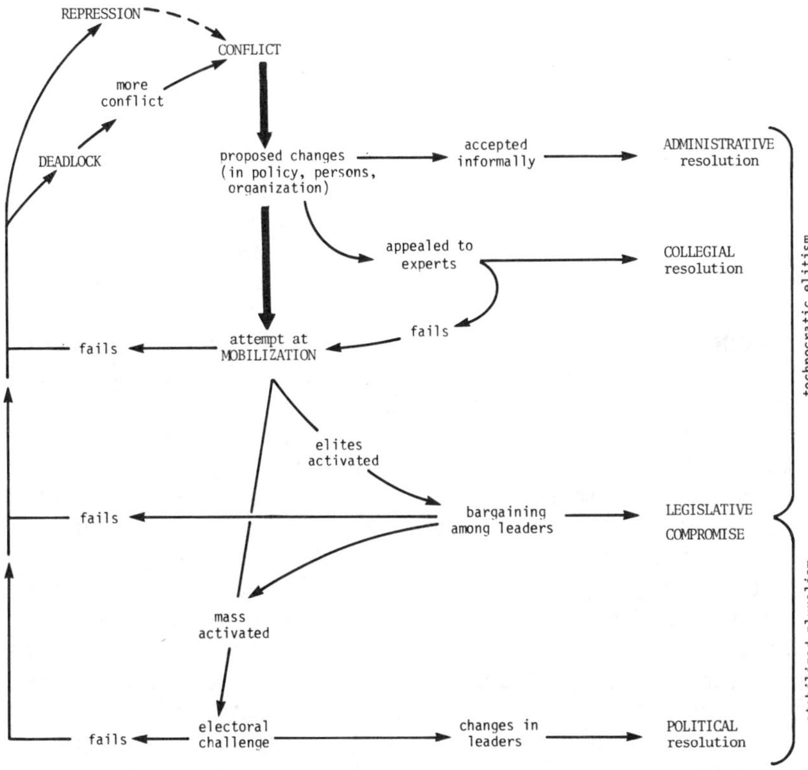

Ideally, a regional governance structure would be accountable to its full regional constituency. The domination of the Tennessee Valley Authority by a minority of those it was created to serve is a reminder that mistrust of unchecked power is a sound attitude.[12] Yet regional accountability is hard to achieve, and harder still to sustain.[13] In particular, the possibility that the electorate could maintain a level of knowledgeable concern sufficient to guide regional power planning seems implausible.

Accordingly, institutional design must face the uncomfortable possibility that electric power will again become an issue of low visibility to the general populace, that it will become either noncontroversial or perhaps worse, the subject of simplistic political posturing. This does not mean that region-wide political accountability is unimportant, only that it will be difficult to achieve and to sustain. Some guidance comes from the work of the sociologist Philip Selznick, whose landmark study of the Tennessee Valley Authority insightfully probed the process of institutionalization.[14] In his later work, *Leadership in Administration*, Selznick proposed three essential ingredients shared by all successful institutions.[15] The first is a set of institutional objectives and purposes, sufficient to engage the loyalties of organization members--an institutional ideology. Second, the support of external constituencies, such as the utility and industrial support that undergirds the iron triangle of BPA. Third, successful institutions must endure, and they must accordingly make provision for recruiting new leadership as time passes.

Selznick warned that institutions can lose their institutional mantle through an erosion of the values that guide them. The critical failure of leadership, he suggested, is default--a failure to understand that the organization must be seen to have purposes larger than its official mission, if the esprit of its members is to be renewed and sustained.[16] A common mode of default, he observed, is the retreat to technology: organizational leaders concentrate on solving problems that can be solved through well-known procedures, rather than meeting the challenges that threaten the viability of the institutional community.[17] The retreat to technology has been a prominent feature of Technocratic Elitism--and the inadequacy of this approach is the primary reason why the PNUCC bill failed once it became controversial.

Nonetheless, the default of leadership does not elicit swift challenge, especially if leaders command sufficient authority. "By setting the limits within which policy alternatives are to be selected," Greenstone and Peterson comment, "the ideologies and specific preferences of the authorities do not constitute just one element in the structure of power. Very often they are the single most important link between the distribution of resources and the final decisions."[18] Although effective in the short run, insensitive use of authority has a repressive effect (Fig. 19) and, as Selznick suggests, can undermine the value structure of administrative organizations as well.

Some of the purposes of any institution will be endangered, either because they are controversial or alternatively because they are ignored and accorded low esteem. These Selznick calls "precarious values." An important element of administrative leadership is the nurturing of precarious values; common strategy is to assign responsibility to protect these values to powerful elements of the organization.[19] For example, if a commitment to energy conservation is both precarious and important to the long-term survival of Bonneville, the administrator would be well advised to create an assistant administrator for conservation, rather than having the conservation program report to the power manager, as is currently the case.[20] To be sure, this strategy of protecting precarious values is often more symbolic than real, as one sees in important titles associated with affirmative action and environmental quality, titles lacking resources or political clout in many cases.

Institutionalization is the process of endowing an organizational entity with social values. But as Charles Perrow has observed, it is important to ask where those values come from.[21] Institutions whose ends are sustained by internal values can be effective but unresponsive; the old Atomic Energy Commission was an example. Given the current controversy surrounding regional power, it is clear that regional leadership must reinstitute the "public cognizance,"[22] and draw upon values that range far beyond those of the Technocratic Elite. It is no longer enough to manage the regional power supply system by its own internal logic, as the Hydro-Thermal Program attempted to do.

Although the public power ideology of the New Deal is unlikely to be revived, it has historically provided political coherence: a sense of purpose grounded in public values. This remains critical in two respects. First, a renewed vision of what the regional power supply system is for would help to stem the tide toward single-issue special interest politics which seems increasingly to characterize the American polity.[23] Second, of independent importance, an institutional ideology is indispensable to utilities and those interest groups which support and criticize them. The sense of responsiveness of utilities to their consumers, damaged by environmental controversy and rapid rises in retail power prices, must be restored. The consumers of the region need to be able to trust their utility officials once more, as the Northwest heads into a period where the reliability of power supplies will of necessity be less certain than it was when federal surpluses were the order of the day.

It is therefore crucial to institutional design that the public interests to be defined in the Northwest are numerous and mutually incompatible: short-run economic impacts can only be mitigated by lowering or stabilizing prices--weakening the economic pressure for conservation. Therefore, *more than one institutional voice is needed*. That is, no single organization should control regional power affairs. Instead, a network of interdependent organizations is required, to articulate publicly the competing views that need to be balanced and re-balanced in the task of regional governance.[24]

In particular, state governments are not about to surrender their authorities in energy facility siting, regulation of retail rates for electricity, or the police powers needed to enforce conservation policies to any multi-state body. And government at any level will not take over the building of thermal plants, nor attempt to legislate dramatic conservation measures, nor suspend Constitutional rights to property or even to protest. Many of the powers that could arguably be "needed" by a single regional energy authority are simply not available. Regional governance must involve shared--that is, intergovernmental--leadership and institutional redefinition.

A persistent theme of this new institutional structure will be the conflict which has plagued the existing one: competition between short-run economic interest and long-run objectives. The rate disparity issue in Oregon, Idaho, and Montana has generated public awareness that short-run interests are at stake in regional power planning--but without any accompanying sense of the long-term reverberations of those interests. This foreshortening of the intricacies of utility planning into the controversy over hydro allocation can only make institutional maintenance itself precarious.[25] Even if parochial economic interests remain primary, however, there will be considerable tension with less immediate objectives.[26]

Despite the steep rise in electric power rates, energy conservation, environmental quality, and an enhanced sense of the social ends served by energy are all objectives which cannot be removed from the agenda, nor wholly sacrificed to short-run gain.

The challenge facing the leadership of regional institutions is to develop programs able to meet the challenge of these goals in a way that is politically compatible with the shared concern to protect vested economic interests. The structure of these goals, as Haas suggests, should be somewhat inconsistent:

> Planning should not be wholly rational. It must be manipulated politically so as to find tasks that will generate enough controversy to lead to new tasks but not enough controversy to endanger organizational survival.[27]

The danger today seems to be too much rather than too little controversy, although the rapid decline of Bonneville's leadership in public power after the Second World War is a warning that too little controversy can be equally dangerous to the public purposes served by governmental bureaus.[28] (Lest the tone of Haas' dictum seem uncomfortably Machiavellian, one should bear in mind that his concern is for the institutionalization of precarious values. In a situation of great complexity, moreover, standardized rules will fail because the system to be managed is both changing and diverse; here conflict and controversy act as corrective signals that are part of the steering mechanism.)[29]

One should not underestimate the magnitude of the change involved. The iron triangle based upon short-run economic interests survived presidential imposition of the partnership policy; the realignment suggested in Figure 22 is at least as profound a change. Indeed, it is precisely for this reason that institutionalization is important--social forces larger than those available through reorganization must be mobilized. In particular, ties to interests and interest groups that have

Figure 22. *The Iron Triangle Renovated.*

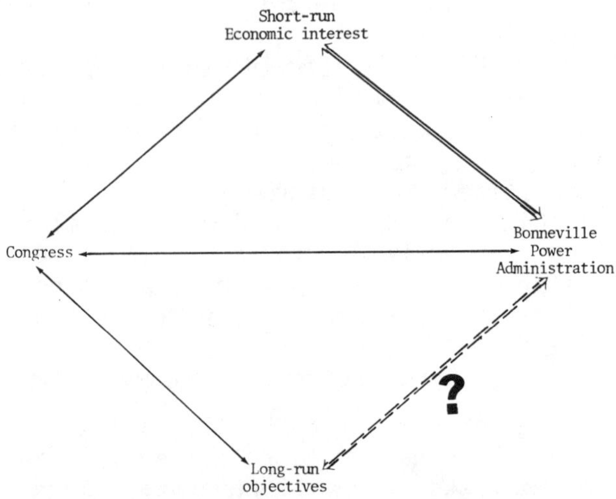

not been represented in the past need to be developed. The rising concern for energy in state and local governments is accordingly an important opportunity to be grasped here. Moreover, following Selznick's suggestion, a revised "iron quadrangle" must provide the opportunity for new leaders to be trained and selected.[30]

POWER AND ACCOUNTABILITY

Institutionalization of broad public values in the regional power program is the central element of managing the conflicts which lie ahead. Yet regional authority is at best a fragile idea in the American governmental tradition.[31] Moreover, the legacy of public power means that the very specification of which responsibilities are governmental and which private is continuingly contentious.[32] Regional governance must thus be grounded on that set of tasks which is genuinely regional, and focus upon responsibilities that are crucially public. Wise policy does not mean more government, nor does it necessarily mean a new layer of government at the regional level. Neither, in the late twentieth century, is likely to be workable.

However Congress may alter the legal mandate, the six regional challenges identified in Chapter V--new generation; financing; allocation of hydropower; energy conservation; reserves; and governmental arrangements--will remain. Work on these challenges does not, of course, begin with a clean slate. There is already a regional power system, and a regional organization managing it. While there is no need simply to endorse the traditions and precedents embodied in BPA's organization and practices, the mere disbanding of Bonneville would only create the urgent need to reinvent it.

The salience of power planning today, however, makes it possible to evolve a definition of the public interests to be nurtured, a definition reflecting broad democratic judgment.[33] Such a near-term definition of the "public cognizance" would comprise the core of the organizational ideology needed to achieve successful institutionalization of regional organizations, and thus flesh out the bare bones of new legislation.[34]

We discuss next two rather different mechanisms for enhancing institutionalization. The first is an elaboration of the regional council proposal developed by the governors of the four Northwest states as an amendment to the Jackson bill. Next, we shall discuss a large (and probably politically infeasible) change in the preference policy. These proposals run the gamut from the politically acceptable to the far out: we put them forward here because in our view they can move the regional power debate in constructive and sometimes novel directions.

The Bonneville Advisory Council

Institutionalization of broad public values in the regional power program has become an essential element of regional power planning. This is nonetheless consistent with centralizing decision-making authority in the Bonneville Power Administrator, along the lines in the Jackson bill. Such authority would be a natural, even necessary adjunct to BPA's existing operations, given purchase authority and responsibilities in enforcing and financing of region-wide energy conservation.

Decision making should be centralized, but its value base should come from the community served. *The informal and formal basis on which the administrator reaches decisions should reflect a far broader constituency and set of public purposes than has been traditional.* A single Bonneville Advisory Council (BAC), with influence mandated by statutory language, could provide advice and counsel on six central questions of regional power policy:

1. *Financing of investment* in power supply *and* in conservation. Advice on major projects and programs should be provided by the BAC, even though it would often be far from unanimous.

2. *Forecasts and regional plans* developed by BPA should be subjected to annual review by the council, with the assistance of a small staff independent of Bonneville. The BAC should transmit a summary of its critique to the Department of Energy, the Office of Management and Budget, and the appropriations committees of Congress, when the administrator submits his annual budget request. In addition, the governors and utilities of the region should be provided the full critique.

3. *Siting and generating technologies* should be reviewed on a regional basis. State governors should specifically concur with regionally significant sites located within their jurisdictions. Although states are unlikely to yield their siting authorities, the regional plan should draw upon generic reviews of sites and technologies sponsored by the BAC.

4. *Regionwide conservation standards,* appropriately tailored to variations in geography and the differences among residential, business, and agricultural uses of electricity, can be recommended to state legislatures and local governments and to BPA for inclusion in wholesale power contracts.

5. *Appraisal of the long-term impacts of power supply and consumption,* especially the

inherent competition between hydropower and the other uses of the Columbia River and its tributaries, should be reviewed in a series of special studies by scientific panels directed by the council.

6. *Allocation of federal hydropower* should be reviewed by an allocation committee of the council, as discussed below. The reviews should be conducted in accord with criteria proposed by the whole council, and approved by the Department of Energy after publication in the *Federal Register* and regional hearings.

The BAC should be assigned statutory responsibility (1) to advise the administrator with respect to these matters of regional concern, and (2) to develop a public record of region-wide comment on regional power policy and decisions. The administrator should, in turn, be required by statute to state *publicly* his reasons for disagreeing with council recommendations. In addition, the BAC should be given statutory authority to communicate directly with Congress, state and local government, the Office of Management and Budget, and the Department of Energy. In a sense, the Bonneville Advisory Council is the analogue of the military Joint Chiefs of Staff, representing those affected by regional power policy, yet not exercising decisional authority.

The composition of the council should reflect regional interests, as distinguished from federal ones. Moreover, given the diversity of concerns within the region, the means of selection should reflect different configurations of interests: some geographic, some functional, some diffuse and fragile.

It should suffice to have a part-time council, with a full-time chairperson. In order to promote maximum leverage for the council's advice to the administrator, it would be sensible to have the administrator nominate a chairperson once the BAC has been selected; the council could, in turn, refuse the administrator's nomination.

Length of service should reflect the enormous complexity of regional power issues, but also provide for turnover. An appropriate balance--though the numbers are somewhat arbitrary--is for council members to serve five-year nonrenewable terms, with a written commitment to serve at least two years at the time of appointment.[35]

It is indispensable to keep the council small enough to be workable, although the number of interests to be served dictates a substantial roster. A suitable if not sacred design is a council of fifteen members divided as follows (see Fig. 23):

1. The four governors of Idaho, Montana, Oregon, and Washington. Despite the past weakness of state participation, the political legitimacy offered by gubernatorial membership is important and useful. It would be appropriate, moreover, to give the governors extraordinary representation in matters affecting the welfare of their states: for instance, investment decisions in power plants and conservation programs might require concurrence by three of the four governors, including concurrence by the governor whose state is affected by a siting choice made at the regional level.

2. Four representatives of legislatures and local governments, with three of these four to come from Washington and Oregon. The principle of active local governmental participation is important. Consistent with this premise, substantial flexibility could be left to the states.

3. Three representatives of the utility industry, direct-service industrial customers, and major business corporations. The scope of representation for business is necessarily broad, especially because the position of the DSIs in the Jackson bill makes it logical to consider them together with other large industrial users of electric power. We see no reason to exclude public or consumer owned utilities from representation under both this heading and that of state and local government, if both constituencies agree.

Figure 23. *The Bonneville Advisory Council.*

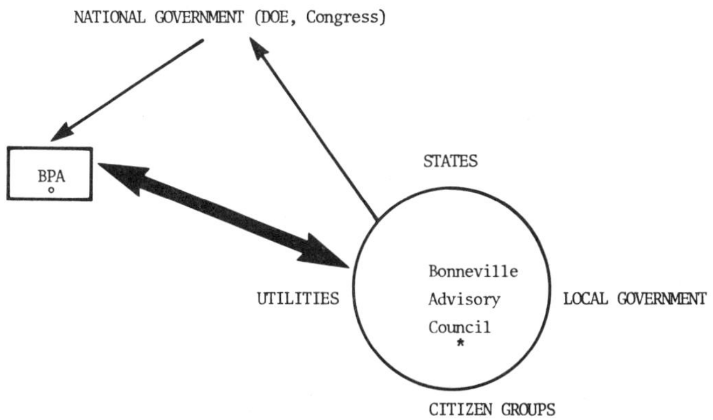

°prepares regional forecast, including conservation and new generation	*critiques regional program
°allocates federal hydro	*recommends hydro allocation; or supervises regional wholesale market
°operates regional grid and FCRPS	*studies long-range implications of federal hydro operation, especially multiple-use conflicts
°assists financing of regional conservation, renewable resources, and thermal generation	*governors concur in new plants

4. Two environmental representatives, selected by the Washington Environmental Council, the Oregon Environmental Council, and analogous multi-organizational groups in Idaho and Montana.

5. Two representatives of non-environmental citizens' groups, selected by the League of Women Voters and other public interest groups, such as the Municipal League of Seattle.

A council structured in this fashion would have a natural tendency to reflect the population density of the Northwest, and yet be freed in important senses from the geographical straitjacket of representation according to rigidly defined jurisdictions. At the same time, the numbers suggested above should not be interpreted as a weighting of the relative importance of interests.

Substantial adjustment is possible, although all the interests specifically named should be represented in the BAC.

The work of the BAC should be supported by a small staff, able to do studies and reviews related to regional power planning. The staff should be composed of professionals with training in economics, utility management, engineering, environmental science, law, and policy analysis. A number of persons should be detailed to serve on the BAC staff from state and local government agencies. Analysts with experience at the Oregon Department of Energy, for example, could give the council a running start in commenting upon and analyzing regional power plans. The size of the staff should be kept small--probably around a dozen professionals. It should function as a "central" staff--responding to the requests of the council or an executive committee of the BAC, rather than to individual members directly. Bonneville should supply administrative support for personnel, administrative matters, and clerical and computational assistance. This would dictate location of BAC offices in Portland--which is probably sensible in any case.

BAC's operational budget, however, must be under its own control. Whether such a budget should be derived--as suggested by the Seattle City Council--by a surcharge on wholesale power sales, or by some other means, the important point is that it be "fenced" as an item in the BPA budget, not subject to control by the administrator.

In most respects, the advisory council we propose can be created without new legislation, under the provisions of Section 655 of the Department of Energy's organic act.[36] We believe it desirable, nonetheless, to have the BAC explicitly chartered by Congress. Such action would provide clearly for the strong powers we think necessary to save the BAC from the ineffectiveness of most advisory boards. Moreover, a clear-cut congressional mandate protecting the independence of council staff and budget is needed as a hedge against future weakening of the council--a potential which would weaken its determination from the outset.

The Jackson bill passed by the Senate designates a five-member council that is at once weaker and more broad-ranging in its authority than the BAC. The bill's

Electric Power Planning Council is weaker in the sense that its political base depends solely on the governors and BPA, and the independence of its staff and budget are not clearly articulated. At the same time, the council's actions bind the Bonneville administrator more tightly, though the administrator retains formal veto power on important decisions. The Senate bill nonetheless provides many of the strengths of the council we propose here.

The functions to be served by the Bonneville Advisory Council are part of the larger institutional structure of regional power operations and policy. The BAC is a consultative body, serving many of the purposes of a regional energy legislature, but without authority on its own to enact policy. Its legislative function reflects the continuing important divisions in the region about the ends which should be pursued by electric power. (See Fig. 20). The compromises endorsed by the council are subject to electoral test in the sense that a majority of the members are elected public officials who serve by virtue of their offices. (This is one reason to include state legislatures and local governments.) The ability to crystallize and to resolve disputes via legislative and electoral means (Fig. 21) permits the BAC to act with the authority--derived from its accountability--of a Stabilized Pluralist model.

At the same time, the inevitable disputes over the means to these ends, including debate over estimates to be used in planning for power and conservation, are structured in an orderly fashion in accord with the spirit--and minimal requirements--of the Technocratic Elite (See Fig. 21). The Bonneville Power Administrator is a regional executive, with decision-making authority that is, in principle, broader than that of any other single agent. In practice, however, the strong advisory powers we would assign to the BAC provide "checks and balances" of considerable strength--if the council develops and exercises its potential influence.

Yet the fact remains that structural redesign by itself cannot solve the problems of regional power policy. "The design of [a policy's] political system, rather than its organizational structure, is the critical consideration."[37] The key to the existing political environment is the set of material interests linking together the Technocratic Elite. If there were a way to tie precarious values such as long-range planning, energy conservation, and environmental quality to decisions concerning electric power, informal patterns of influence might develop which would institutionalize these values. The Jackson bill goes part way down this path, by requiring that cost-effective conservation and renewable resources be developed before thermal resources are acquired by the administrator. This framework ties precarious values to BPA purchase authority. Whether these congressional priorities will be heeded by Bonneville remains to be seen, especially in light of the enormous pressure from the public utilities to spread the costs of the two publicly sponsored nuclear plants not backed by net billing.

Moreover, the vaguely defined regional resource plan and conservation standards in the Jackson bill invite technocratic misapplication. Both these provisions can easily be turned into symbolic defenses of business as usual, thus encouraging cynicism and further opposition.[38]

It may be possible in short, to harness electric power to the cause of institutionalization more directly. This is the purpose of the proposal we take up next.

Incentives Instead of Preference?

Since 1906, federal law has provided that publicly owned utilities should enjoy preferential access to electricity developed at federal projects. The Bonneville Project Act declared that "the administrator shall at all times...give preference and priority to public bodies and cooperatives."[39] With the rising rates experienced by customers of investor-owned utilities, there has been pressure to reallocate the economic benefits of low-cost federal hydro. The Jackson bill provides the residential consumers of IOUs a wholesale power rate equal to that of the public utilities. The question of preference has thus been framed this way: should the benefits of federal hydropower continue to be allocated to public power agencies, or should they be shifted to residential consumers no matter who delivers their power? Both visions of preference are too narrow, for neither set of claimants has bothered to articulate the ends to be served by the public investment represented by federal hydro.

The historical right of public bodies and cooperatives to federally developed power is clear: the New Deal Congress and President Franklin D. Roosevelt saw that multiple-purpose dams would provide broad, continuing economic and social benefits to large regions of the nation. The Tennessee and Columbia Valleys have been the principal beneficiaries of that vision. Basic to this strategy of multi-purpose development was the extension of electric power to rural areas and to the public agencies that were then expanding to meet urban demand.

Since the withdrawal of federal leadership signaled by the Eisenhower Administration's "partnership" policy, the publics have drifted away from their commitment to social progress. This is apparent in the quiet reluctance of Northwest public agencies to assist in the formation of new public entities, now that federal hydro is in short supply. "The publics, some of them, would love to see more public systems," former BPA administrator Don Hodel said. "But with more public systems their prices would go up faster. The Public Power Council executive committee came to me one time and said I should promise new public systems that BPA would meet some proportional share of their load. I said...if you want a new public system formed, why don't you just get together and agree how much each of you will give to them and I can tell you now, we will approve the arrangement. 'Oh,' they said, 'we couldn't do that.' ...I never heard another word about that. That's the real world."[40]

Moreover, the participation of the region's public agencies in large central-station projects, such as the nuclear plants of the Washington Public Power Supply System, has drawn public utilities away from the decentralized, responsive relationship to their ratepayers that constituted the historic strength of public power. And the purposes supported by public power are often parochial and not in the broader regional interest; local irrigation development has been promoted, for instance, at the expense of fisheries and downstream power generation.

If the claim of the public systems is tarnished, that of residential consumers promises at least to be durable. But if public preference is simply replaced by a residential preference policy, the economic discontent of residential consumers served by private utilities would be stilled, and with it the reviving interest in public power. This would be a sad and important loss. However weakened its conscience, public power remains a useful, sometimes influential counterweight to investor-owned companies. Diversity and the competition of the so-called "yardstick" has served the Northwest and the nation well. We believe it possible to modernize the measuring power of the yardstick without discarding it.

Federal hydropower can provide an invaluable incentive for public utilities to serve both their ratepayers and the region as a whole in the tradition of their distinguished history. But the permanent subsidy embodied in the preference clause needs reinterpretation if this incentive is to be harnessed.

Such a dramatic change is needed because the ends to be served by federal resources have become both complex and dynamic. Where growth was once an unquestioned good, it must now be complemented by conservation. Where costs of production once fell before the advance of technology, rising rates now worsen the hardship of the poor. And where industrialization, irrigation, and suburbanization once were the hallmarks of improving quality of life, the agenda of progress now requires informed judgments of subtlety and balance. An appropriate name for the objectives to be pursued is "multiple use," the traditional commitment to balancing the variety of benefits of the Columbia River. Today, multiple use must include conservation of electricity, but in other respects the long-term interests of the Pacific Northwest are still those that guided the New Deal: wise use of resources for economic production and the pursuit of social justice, including intelligent enhancement of environmental quality.

The enormously valuable legacy of federal hydropower can contribute vitally to the search for this more complex, richer public interest--if the benefits of federal hydro are not permanently assigned to either a class of utilities or a class of ultimate consumers.

In place of a preference policy, then, we propose an incentive policy: the assignment of federal hydropower to a Pacific Northwest Public Incentive Pool (PIP).

At the outset, PIP resources would be assigned as they are now, to publicly owned utilities and cooperatives. Every three years, however, each utility receiving PIP would be required to justify its access to the benefits of PIP. Utilities would submit reports to an allocation committee of the Bonneville Advisory Council; the committee's membership should by statute include representatives of the utilities, and consumer and environmental groups.

Reports to the committee would describe the ends served by PIP in the preceding three-year period. Among the broad public purposes that should be specified by Congress are energy conservation, social justice, protection of natural resources, and business development that improves equality and employment. In addition, a utility might use Public Incentive resources to advance a regional or national policy objective, such as petroleum conservation through electrically assisted solar space and water heating. These reports would be subject to critiques by citizens and business and governmental organizations.

If in the judgment of the allocation committee a utility's use of Public Incentive resources met the evolving standards of the public interest, that utility would retain its allocation. If not, 15 percent of its initial grant of PIP power would be withdrawn; the utility would be given access, on a three-year contract, to wholesale power at the prevailing rate. The difference between the prevailing cost of power and PIP prices would be used to lower the prices of all classes of Bonneville customers. The withdrawal of PIP resources is equivalent to a surcharge on 15 percent of the allocation. The magnitude of the surcharge is set by the prevailing cost of non-federal power. In contrast to the Jackson bill's provisions, our proposal does not use withdrawal of PIP as a punitive device, but signals the market-determined cost of providing electric power. After another three years, a utility that had lost part of its PIP allocation could apply for reinstatement by submitting a report describing how the utility, its ratepayers, and its local government had utilized their total electric power sales in support of the public interest.

The recertification procedure resembles that required of holders of broadcast licenses from the Federal Communications Commission. We are well aware that the FCC has often been criticized for insufficient vigilance in its regulatory undertakings; this is, by analogy, a potential weakness in our reallocation proposal. We deem it an acceptable one, however, since mere inaction by the allocation committee would result in preservation of the allocations defined initially by congressional policy.

By requiring review every three years, the large number of utilities in the Northwest could be divided into three manageable pools, one pool to be reviewed each year. More than 2,000 average megawatts of low-cost resources would be available for reallocation each year--a significant incentive in its own right for consumers seeking to scrutinize the reports of utilities under review.

The judgments of the allocation committee would, over time, form a body of "case law," serving to define the moving target of the public interest. As this judicial language suggests, precedent should influence the allocation committee heavily, so as to provide predictability for utility planning. A utility that ignored the regional interest so defined would receive a gradual but strong economic signal; over a twenty-year period, a utility consistently pursuing a selfish course would lose all PIP benefits. Examples of "selfish" behavior include pursuit of or acquiescence in load growth without compensating conservation programs; soliciting of new commercial or industrial customers without accompanying attempts to ensure that service-area residents gain additional employment; or refusal to consider step-up tailblock rates.

At the same time, the allocation committee of Bonneville Advisory Council acquires "legislative oversight" functions through PIP review, with the 15 percent withdrawal playing the role of a cut in appropriations by Congress. Such leverage would strengthen the BAC dramatically by vesting it with the material leverage of economic sanctions. Without such significant power, the BAC would be hard pressed to prevail against the

Technocratic Elite. "The tendency of democratic participation to break down into administrative involvement," Selznick warned, "requires continuous attention....The things which are important in the analysis of democracy are those which bind good men....Something more than virtue is necessary in the realm of...power.[41] It is appropriate to place resources for "something more than virtue" in the element of our institutional design where public accountability is concentrated.

The definition of the public interest may seem perilously vague, and the judging of it subjective. Two recent analyses have posed the problem clearly:

> The single most important characteristic of the newer forms of social intervention is that their success depends on affecting...hundreds of millions of individuals, millions of business firms, and thousands of local units of government. The tasks are different, not so much because they deal with technologically complicated matters as because they aim ultimately at modifying the behavior of private producers and consumers. The boundaries of the "public administration" problem have leapt far beyond the question of how to effectively organize and run a public institution and now encompass the far more vexing question of how to change some aspect of the behavior of a whole society.[42]

> What our society requires most urgently is not a corps of judges earnestly striving to do what is "good"; what we require instead is to recognize the defining "the good" in an industrialized society is a complex task.[43]

These are serious challenges: can the Bonneville Advisory Council guide millions of consumers' judgments deftly enough to avoid the resistance that environmental regulation has engendered? And if it can, will the public interests to be pursued be at once clear enough to implement and flexible enough to adapt to uncertainties ahead? Each question requires a careful comparison of the PIP mechanism to the alternatives set forth in the Jackson bill.

Note first that our proposal affects the prices paid for power. The PIP process thus utilizes economic signals to "inform" consumers of the public purposes to be served by federal power. We put "inform" in quotes to emphasize that prices cannot transmit subtle messages. At the same time, the effectiveness of price signals depends on their simplicity. PIP is thus similar to the effluent charge, a means of charging for the right to dump pollutants into the environment.[44] More broadly, PIP belongs to a set of policy instruments that use market mechanisms to allocate scarce "public goods"[45] which cannot be sold efficiently through the free market.[46]

Second, it is worth pointing out that there are straightforward means of establishing criteria in each of the areas we recommend for inclusion in the regional public interest. For example:

1. *Conservation* of energy in a manner that is *cost-effective*, when measured against the marginal delivered costs (both generation *and* transmission) of power

2. *Social justice* as realized in *conservation subsidies* to the poor, combined with *rising tailblock* rate structures, so that those consuming large quantities of energy pay their (disproportionately larger) share of the cost of electricity

3. *Conservation of natural resources* by protecting *Columbia River salmon* through management of peak power demands on the hydropower system[47]

4. *Sensible business development* that provides *more jobs per kilowatt-hour* consumed in production than the regional average.

These are not the only criteria that could be established by the BAC. But they are clear, quantifiable, and subject to independent examination and verification at reasonable or even minimal cost.

Cost-effective conservation is probably the most immediately important of these regional objectives. Here, one should note that David Bardin, Economic Regulatory Administrator of the U.S. Department of Energy, stated before the House Subcommittee on Energy and Power that "the administrator of the Bonneville Power Administration should be specifically authorized to require, *as a condition of wholesale contracts,* that distribution utilities institute comprehensive conservation programs, in cooperation with state and local governments."[48] The spirit of this recommendation is close to our own: low-cost federal hydro should be bent to the pursuit of public purposes. We would go beyond conservation, and we would modify the implementing mechanism of Bardin's recommendation by giving the regional voice of the Bonneville Advisory Council weight commensurate to that of the federal Bonneville administrator.

If Congress were to validate the claims of residential consumers now served by investor-owned utilities, the Jackson bill provides a mechanism for providing them the economic benefits of federal hydro. Residential customers could be given an initial allocation of PIP resources under the ground rules we propose. In the near term, this could be done with no loss to the existing publicly owned utilities. In the longer run, it is less clear how investor-owned utilities should justify their customers' continued access to PIP power. The special difficulty is that most IOU service areas cover several governmental jurisdictions. Much of the time, serving the public interest requires cooperation among utility, ratepayers, and local government. Should the failure by an IOU to secure beneficial local ordinances in part of its service area be grounds for withdrawing PIP resources from all customers served by that utility? The "case law" to cover such situations would be more complex, though there do not seem to be any insuperable difficulties.

Such an allocation procedure would reinvigorate the public power movement. At the same time, the competition for the economic benefits of Public Incentive resources would serve as a powerful incentive to all utilities to align their practices with the broad interests of the region. (It is worth pointing out that the Jackson bill can make use of a PIP allocation structure, with the regional council serving the role of our allocation committee. The unconventional notion of using federal hydro as an incentive is unlikely to be implemented, however, without explicit congressional sanction.)

So profound a revision of the historic policy of public preference clearly raises a host of legal, administrative, and political questions. We consider here some of the more obvious ones.

An important complication is the definition of the Public Incentive Pool. Power generated in federal dams is mostly committed under twenty-year contracts to publicly owned utilities and direct-service industrial customers. The costs of four thermal projects are mixed into the federal pool. These nuclear projects are financed through net billing. Only one partially net-billed plant, the Trojan nuclear project in Oregon, has gone on-line. The wholesale costs of federal hydro and net-billed power are expected to rise rapidly, although they are likely to remain lower than the average wholesale costs of non-federal power. The magnitude of the incentive in our Public Incentive resource would nonetheless be diminished as wholesale costs rise.

We believe that the Public Incentive Pool should include *only* federal hydropower. Net-billed thermal costs should be separately allocated. A similar rate pool definition was advocated by the Snohomish Public Utility District in the amendment it prepared to the PNUCC bill in 1977.

A second challenge to our design is the claim that the allocation committee of the Bonneville Advisory Council could not function effectively. Its burden is a heavy one, even with the reliance on precedent and the traditions of public power that we suggest.[49] The usual protection of judicial independence is insulation from political pressures through long tenure; we regard this as an inappropriate solution, given the rapidity with which regional power issues may evolve. In the end, there is no satisfactory solution short of appointing independent, determined, intelligent committee members. In its early years, the Supreme Court moved decisively to define its role; a similar challenge awaits the allocation committee.

There are a number of technical details that significantly affect the legislative language implementing our proposal. The most visible is the need to exempt publicly financed generating plants from the provisions of the Revenue and Expenditure Act of 1968, as interpreted by the Treasury in 1972. These regulations set a limit of 25 percent on the total output of a publicly sponsored power plant which may be sold to "non-exempt persons," including BPA and the investor-owned utilities. In order to achieve the power exchanges needed under the reallocation provisions of our proposal, the tax exemption of the public sponsors of regional facilities should be expanded to 100 percent of the output, lest third parties be injured in the pursuit of public interest incentives.

The strongest barrier in the way of our proposal is political acceptability. We believe that our proposal adapts the policy of public preference in a manner suited to the circumstances of the late twentieth century; it provides a sound basis for preferring utilities which serve the public good. In that sense, the "yardstick" concept of the New Deal is given a more complex, but serviceable meaning. Most of all, it provides a framework in which the rationale of preference can be articulated fully, apart from the immediate pressures of rate disparities reflecting ephemeral differences in retail rates.

OPTIONS IN REGIONAL GOVERNANCE

Our analysis has led from design premises to organizational missions with the potential for fulfilling the political as well as managerial needs of institutional design. Further growth of the regional power supply system raises questions of enormous public significance, far beyond the provision of electric power. The choice of institutional arrangements must not be based solely on narrow considerations of electrical technology and economics. Instead, we should look to the structural arrangements that can best assure consideration of long-term implications and the broad public interest. Sadly, society's ability to engineer organizations to achieve these goals is not equal to the challenge.

We have mentioned the difficulties of assuring public accountability at the regional scale. No elected official now has a regional constituency. We doubt the electorate can sustain--or even achieve--a level of knowledgeable concern sufficient to provide meaningful guidance via direct election of regional energy commissioners. The history of commissions appointed by state governors is not hopeful either, as can be seen in the performance of the Delaware River Basin Commission.[50] For years, we have relied upon the responsiveness of Bonneville to the Northwest congressional delegation and especially Senator Jackson, who has been influential as chairman of the Senate Interior and Energy Committees. Can the region now go beyond the personal ties of the Technocratic Elite? Let us consider the span of the feasible. How narrow that span is in fact may be measured by three institutional alternatives.

1. *Continued deadlock*. The primacy of short-run economic interests makes continued deadlock far from the least likely alternative. Although the course of turmoil is hard to predict, it is possible to identify deadlines which, once passed, will have objective consequences for the region.

The BPA Notices of Insufficiency, issued in mid-1976, are projected to fall due in 1983, and Bonneville has recently proposed an allocation beyond that time, should no new policy framework be in place.[51] Of equal consequence, a continuing stalemate dividing public and private power would probably worsen relations between Washington--where public power is centered--and the other Northwest states. Such a development would put to a severe test the regional perspective that has consistently been pressed by Senator Jackson.

Prolonged deadlock would probably lead several direct-service industries to abandon the Northwest as the time came for major capital additions to their facilities, additions that could not be justified in view of the continuing uncertainty of power supply and prices. While this would produce outcries that jobs were being lost, the greater cost to the region might be the loss of the economically productive use of reserve capacity that is the historical contribution of the DSIs.

Eventually, the go-it-alone strategy forced upon the investor-owned utilities since 1973 would come to characterize all the utilities, with public agencies drawing upon the power "liberated" by the preference clause as DSI contracts are not renewed, but eventually relying upon thermal power consortia like WPPSS. Under these circumstances, the preference policy would maintain the disparity in wholesale power rates that is a principal incentive for institutional redesign; one estimate made by BPA projects that by 1995 retail customers served by IOUs will pay rates 275 percent higher than those served by preference utilities.[52]

Even before that time, BPA's manifest inability to resolve a deadlock would provide a ripe opportunity for plausible rivals for regional leadership, notably WPPSS. An indication that this logic is persuasive may be found in an "Assessment of Arrangements" in the event no legislation is enacted, prepared by WPPSS in April 1978.[53] WPPSS called for an organizational structure paralleling the existing West Group, with IOUs removed, and with a new Public Pooling Agency serving public utilities and DSIs. The Public Pooling Agency--a joint operating agency that is clearly meant to be WPPSS--is a replacement for BPA, performing Bonneville's functions for the DSIs and the publics, who would assign their federal hydro entitlements to the Pooling Agency. There is scant indication, however, that Bonneville is interested in being superseded, and a BPA discussion paper of June 1978 outlines the complex texture of legislative failure from the federal perspective.[54] The important point is that interregnum breeds rivalry.

2. *Defederalization*. The boldest regional vision is one that provides a benchmark of how much institutional forms *could* change. That benchmark is a defederalized Columbia River Power System, with dams, transmission lines, and administrative staff transferred from the Bureau of Reclamation, the Army Corps of Engineers, and Bonneville to some interstate regional authority. Such a reconstituted Columbia River Power System would put into a different context the vexing question of public preference in the allocation of federal hydropower, and it would therefore provide room for institutional redesign.

It is even conceivable that prolonged deadlock might lead to a particularly bizarre form of defederalization: a division of federal hydropower among the several states and a relegation of Bonneville to a passive role as transmission and billing agent.

The Northwest states have diverse interests and aspirations. Oregon's formation of DRPA could be followed by similar legislation in Idaho and Montana, to assert preference rights and a claim to status as some kind of statewide operating agency. This alone is a strong deregionalizing factor, but there are others. The upstream states, Idaho and Montana, are reluctant to provide energy service--whether in the form of upstream hydro storage or in the shape of thermal plants--to the downstream states. Their wariness is matched only by the downstream states' unwillingness to serve California with Northwest energy parks. Idaho attaches high priority to reserving both water and electricity for irrigation and irrigation pumping, and has no aluminum plants. Western Montana, lacking significant irrigation potential, resists dams and strip mining for environmental reasons. As Bonneville's plans for seven additional high capacity transmission corridors across the northern Rockies become public knowledge, the resistance of the upstream states to serving the West Coast will undoubtedly stiffen even more.[55]

The scope of the river basin, the Coordination Agreement, and the Bonneville transmission net may not suffice in the future to overcome the geographic and social diversity of the Pacific Northwest. That diversity has thus far paralyzed all efforts to construct a Columbia River Interstate Compact[56] and any interstate agreement on allocation of water and power. Precedent for direct allocation to the states already exists. In the two controversies over Hells Canyon development, in the 1950s and later in the '60s and '70s, proposals were advanced to allocate power to Idaho. Bonneville power was actually allocated to a new aluminum plant "mandated" in Montana as a *quid pro quo* for the federal Hungry Horse Dam and storage reservoir on the South Fork of the Flathead River. It is a logical, if large, step from allocating some of federal hydropower to one or another state to allocating all of it to the states.

Finally, one should consider the plight of the disinterested congressman or senator from elsewhere in the United States. Given the recently aroused ambitions of the Northwest states to control their own energy destiny, how better to avoid the awkward issues of public preference, federal guaranteed purchase, and the Northwest's seemingly insatiable appetite for new power plants and transmission lines than to turn all the power over to the states and let them make the decisions.

In short, this balkanization scenario may be unlikely, but it is not unthinkable. Because it represents such a major change, defederalization is politically remote. Certainly the preference clause and the financially intertwined irrigation subsidies in the Columbia system would pose enormously complex legal and political challenges to any serious move to defederalize. Though it is mainly a conceptual guidepost, defederalization is a useful measure of the degree to which feasible change is just tinkering.

3. *BPA redux.* Easily the likeliest--and by no means the worst--outcome is the reassertion by Bonneville of its historical role as the preeminent regional energy organization. BPA is the central actor of the Jackson bill, and the agency's functions are, as we argued before, not readily dispensable. The political momentum is there: BPA offers a ready-made organizational frame on which to rebuild the structure of regional power planning. A revitalized Bonneville could proceed with the allocation of federal hydropower; finance new thermal plants and regional conservation measures; reaffirm the regional role of the direct-service industrial customers; and conduct public involvement in region-scale planning.

Most important, BPA would be able to finance new sources of supply on a regional basis. This will inevitably bring changes in the agency's role and responsibilities. Authority to purchase the "capability" of new power stations constitutes a guarantee to repay investors whether or not the plants work satisfactorily. If they do not, higher rates will be charged Northwest consumers to cover the expense of facilities that do not produce as much electric power as planned. This approach shifts the risks of new plant development away from those supplying capital and onto ratepayers; as a result, interest and other financing costs should be lower than they would otherwise be.

Bearing the risk of new resources within the region is a sound idea in principle. It should be matched, however, with vigorous oversight by BPA on behalf of consumers.[57] As our discussion in Chapter IV noted, Bonneville's attempts to exercise these responsibilities for the three net-billed plants being built by WPPSS were rebuffed by the Supply System. Bonneville oversight has done little, in any case, to contain the skyrocketing costs of these projects. An important complement to BPA purchase, accordingly, are safeguards to hold down costs in new Northwest power plants.

Although a mandate to supervise generation projects is contained in the Jackson bill, it could be strengthened in three rather different ways: providing an explicit cost standard in the legislation which, if exceeded, triggers congressional review; permitting states and public utilities to sue to enforce BPA oversight; putting a "sunset" provision in purchase authorization.

The need for a cost standard is evident in the confusion surrounding the issue of whether WPPSS should be subjected to closer scrutiny. A practical method of measuring the performance of projects under construction --a critical period in terms of effective intervention by outside parties--is to evaluate thermal projects against the standards provided by the industry itself.

All thermal projects have experienced sharply escalating costs in the 1970s, as discussed in Chapter IV. It is thus inappropriate to compare a thermal project started in 1980 with one begun in 1975: the cost base and regulatory environment are both different. It is possible, however, to *compare a project launched in 1980 with all other projects begun that year*. A sensible perspective is to match the anticipated cost per average delivered kilowatt at the busbar of a given project against the anticipated cost of all other electrical generating projects which receive initial financing in the same year. In computing this anticipated cost, external social and environmental costs that can be quantified must be included, if a fair comparison is to be made.

Such a moving standard contains some built-in advantages for Northwest utilities: first, the financing benefits in the Jackson bill give regional projects a sizable cost advantage; second, the preponderance of baseload plants needed in the regional supply system will cause the Northwest's utilities to purchase facilities that are, on average, cheaper than the peaking or load-following plants needed elsewhere; third, the advantageous environmental and demographic setting of the region will make external social and environmental costs lower than in other parts of the nation; the abundance of water should considerably ease cooling-system costs, for instance.

For this reason, BPA purchase authority should lead to significantly better performance here in the region than the national average. If it does not, that by itself is significant evidence of mismanagement or other extraordinary circumstances. It would be appropriate, accordingly to trigger a GAO and congressional investigation if the national standard is exceeded.

The cost standard provides information; authorizing states, municipalities, PUDs, and rural coops to sue the Bonneville administrator provides local clout. Such a limited form of "private attorney general" power permits public agencies representing ratepayers to bring suit against BPA for faulty oversight of purchase agreements. A less limited authorization is already in force, under the language in the Bonneville Project Act which directs the administrator to follow "sound business practices." But the legal standard implied by this phrase has provided no remedy for BPA's weaknesses with respect to WPPSS Nuclear Projects 1-3. If a stronger oversight responsibility is needed, authorizing ratepayers' legitimate representatives to seek judicial relief is one way of encouraging it.

Finally, Congress could make purchase authorization clearly experimental by limiting its use to the ten years after passage. Such "sunset" language would limit the ability of the administrator to *enter into* purchase agreements to the period from passage to (for example) 30 September 1990. Since purchase is a contractual agreement, any purchase arrangements in force on that date would not be terminable except on conditions agreed to by both contracting parties. Sunset therefore does not impair the ability of BPA to proceed with a regional power program. Modifications to existing purchase agreements after 1 October 1990 could also be permitted via "grandfather" language limiting changes to (for instance) 30 percent of the dollar value of capital costs projected on 30 September 1990. It would be necessary for Congress to examine the success of the purchase agreements in the late 1980s, and to consider whether to extend--for a limited or unlimited time--BPA's authority. Otherwise, the situation with regard to regional financing of resources would revert to that prevailing today.

An underlying motive for a time-limited purchase arrangement is to encourage the development of alternative regional financing structures. The fact that BPA is today the only plausible regional factor will be strongly ratified by an unlimited purchase authority. But the federal character of Bonneville's role and mission lies at cross-purposes with its regional responsibilities to the Northwest. In the long run, therefore, there may be significant benefits to both the Northwest and the national to defederalizing the regional financing role.

As suggested above, an alternative factoring entity which may become viable within the next several years is the Washington Public Power Supply System. Though WPPSS is now in disfavor because of its management of thermal power projects, in principle a public utility-sponsored financial and construction agent meets the specific needs of the Northwest's utility industry structure well. Indeed, a case can be made that such a public-based mechanism is superior to BPA purchase, with its exemption from the Public Utility Holding Company Act for investor-owned companies. From the standpoint of sound social policy, it is worth encouraging revitalization of entities like WPPSS, even if Bonneville purchase were to be reaffirmed in ten years.

Even without these additions to purchase authorization, the Jackson bill is likely to alter Bonneville's institutional profile profoundly. At its best, the Jackson bill could rejuvenate a broad federal mandate to manage regional power supply. With its responsibility for conservation at the regional level, Bonneville would increasingly be concerned with non-electric energy policies. More broadly, the intertwining of energy policy and social purposes will continue to keep power planning political. The important question is how it can be more than just political, but politically sensitive.

Selznick's analysis suggests that, beyond a coherent set of public purposes, there are two social requirements for institutionalization: support from interest groups and the formation of an elite from which future leaders can be drawn. Financing of thermal plants and regional conservation investments provide linkages to interest groups, through the economic clout they represent. An important element of BPA leadership, then, is to recognize that the tension between thermal plant and conservation investments is a healthy one for the region to face on a continuing basis. This is the basis for recommending that the regional council consider generation and conservation in tandem and explicitly. In particular, sites should be identified for generating facilities and specific utilities should be credited with future conservation savings--together with appropriate economic incentives to achieve those savings.

Regional forecasting provides opportunities to develop knowledgeable critics, from whose ranks future leaders can emerge. The uncertainties of forecasting call for collective judgments drawing upon the growing analytic resources of state and local governments. A Forecasting Review Committee should be created by the regional council, with membership drawn from the utilities, utility regulatory authorities, and state and local executive agencies like the Oregon Department of Energy, to supplement--and to disagree with--the West Group forecast, its methodologies, and its assumptions.

But the uncertainties of the future range beyond forecasting of loads and resources. BPA should also be instructed by Congress to publish periodic revisions to the Role Environmental Impact Statement. These updatings should describe intentions to purchase power from new plants, with a brief statement of their accompanying environmental impact; other program changes affecting regional interests in and related to electric power development and operations; and monitoring summaries, to assay the actual effects of Bonneville programs carried out in fulfilling its role in the region. With such information regularly disclosed, and with the now routine processes of review conducted by other federal and state agencies and interest groups, there is the possibility that information influential in regional power decisions can be regularly developed.[58] This use of the National Environmental Policy Act's impact statement process is a sensible adjunct to the strengthening of BPA, one that makes use of existing mechanisms rather than calling for major structural change. By the same token, of course, the fragility of relying on NEPA as a disclosure mechanism is manifest.[59]

With the institutional alternatives discussed in this chapter, particularly a strong Bonneville Advisory Council, it is possible to put into perspective the role of BPA. The Jackson bill places great reliance on the agency to administer a regional power program, so great that some critics have said that the Bonneville administrator is to become an "energy czar." Yet the historical BPA has been too weak rather than too strong: retreating from its early commitment to public power; acquiescing in the erosion of multiple-use management into single-minded exploitation of hydropower, especially by the mid-Columbia utilities; endorsing the technocratic myopia that confused conservation with "irresponsible" environmental criticism. Given suitable mechanisms for accountability and sound leadership in the next few years, a powerful BPA could establish lasting institutional patterns. The key word is "lasting": effective nurturing of the unusual priority accorded conservation and renewable resources will require staying power. A strong Bonneville can become a "lightning rod," attracting the support and opposition necessary to keeping public issues before the public eye. What is needed is not a "czar" but a "prime minister," an administrator able to govern effectively because he is accountable.

Yet the structural counterweights we have presented may prove unavailing, like any attempt to legislate moral behavior. Bonneville predominance could simply minimize public discussion and the institutionalization of public values in regional power planning. The agency's standing commitment to a technocratic viewpoint has in recent years led Bonneville to be grudgingly responsive at best to concerns beyond those of electric power. Moreover, BPA's traditional devotion to large-scale, centralized power projects does not reassure those who look to unconventional alternatives--and who hope for support from regional power policy. It is encouraging that discussion of a regional conservation program has become considerably more sophisticated in the course of controversy over congressional legislation. But environmentalists are skeptical still.

It is unlikely that the Northwest will return to the pattern of state and local government apathy that prevailed in the hydroelectric era. But people outside the utility circle might have to fend for themselves under BPA dominance. The creation of public bodies such as the strong Bonneville Advisory Council is a hedge against ignoring the broader agenda. It is not a guarantee; given the concentrated powers stemming from purchase authority, there can be few purely structural assurances that the public cognizance will succeed. It can, however, be given a chance.

The agenda of regional power planning is irreducibly institutional. Institutional commitments have historically shaped the positions taken by regional leaders and the outcomes they have fashioned. In particular, the managerial responsibilities of BPA and the utilities have commanding influence over the political stances they adopt and the focus on allocation of federal hydropower which has dominated the agenda. As a consequence, public discussion has yet to engage with the fundamental, long-term choices that face the region.

The Pacific Northwest has often led the nation in experimenting with social reforms. The public power movement was itself a notable embodiment of liberal innovation. The historical situation of the region is, despite the present air of chaos, unusually fortuitous: a strong regional economy; natural resources still sufficiently abundant to provide plentiful environmental amenities in combination with moderate growth; and the willingness--now being tested--to work together as a region toward an articulate public concern for the long-term future.

That future embraces much more than electric power. But it will be profoundly influenced by the institutional and technological dimensions of regional power decisions. We must live with the consequences of these decisions; we face strong pressures to make these decisions without delay. We should make these choices from a regional perspective, with an appreciation of the subtleties of that choice:

> It is instructive to consider how impermanent the greatest achievements of material progress have proved to be....Consider the greatest engineering achievements of the 19th or of the early 20th century: the Suez Canal,

> the American transcontinental railroad, the Trans-Siberian railroad, the Panama Canal. These marvels of engineering have not only changed the face of the earth, they involved the destinies of entire continents. Less than one hundred years after completion what happened to them....? Their engineering followed...what people *thought* was progress: and the future has seldom conformed to the projections of this kind of calculation.[60]

We shall have no more than a year or two to choose for the generation to come.

Notes

Sources and Abbreviations.

Full bibliographic listings of the references used in this work are compiled in a separate Bibliography. Interviews by the authors are referenced by initials (Klemka, DLK; Lee, KNL; Marts, MEM) and interview transcripts or notes are filed at the respective authors' offices.

Several references are used in a number of different chapters, and we have elected to shorten their bulky titles:

BPA Role EIS - Bonneville Power Administration, *The Role of the Bonneville Power Administration in the Pacific Northwest Power Supply System, Including Its Participation in the Hydro-Thermal Power Program,* draft environmental statement, U.S. Department of the Interior, DES 77-21, July 22, 1977, in seven volumes:

Part 1. The Regional Electric Power Supply System
Part 2. The Role of BPA
Appendix A. BPA's Power Resources, Acquisition, Planning, and Operation
Appendix B. BPA's Power Transmission
Appendix C. BPA's Power Marketing
The Alumax Environmental Statement
Summary Report

(The Role EIS is the most comprehensive compilation of data and statistics on the Northwest power supply system.)

Power Review Pacific Northwest River Basins Commission, *Review of Power Planning in the Pacific Northwest, Calendar Year 1977* ([Vancouver, Wa.]: PNRBC, 1978). The 1975 version was published in 1976. (An annual compilation of data relating to power supply planning.)

Senate Hearings — U.S. Senate, *Pacific Northwest Electric Power Supply and Conservation Act*, hearings before the Committee on Energy and Natural Resources, 95th Cong., 2nd Sess., Publication No. 95-154, in two parts (Washington, D.C.: Committee Print, 1979).

House Commerce Hearings — U.S. House of Representatives, *Pacific Northwest Electric Power Issues*, hearings before the Subcommittee on Energy and Power, Committee on Interstate and Foreign Commerce, 95th Cong., 2nd Sess., Serial No. 95-193 (Washington, D.C.: Committee Print, 1979).

House Interior Hearings — U.S. House of Representatives, *Pacific Northwest Electric Power Supply and Conservation*, hearings before the Subcommittee on Water and Power, Committee on Interior and Insular Affairs, 95th Cong., 1st Sess., Serial No. 95-26, in five parts (Washington, D.C.: Committee Print, 1978).

Senate Report — U.S. Senate, *Pacific Northwest Electric Power Planning and Conservation Act*, report No. 96-272, 96th Cong., 1st Sess., July 30, 1979. (To accompany S. 885).

GAO Report — Comptroller General of the United States, *Region at the Crossroads--the Pacific Northwest Search for New Sources of Electric Energy*, report No. EMD-78-76 (Washington, D.C.: U.S. General Accounting Office, August 10, 1978).

NRDC Alternative Scenario — Roger Beers and Terry R. Lash, *Choosing an Electrical Energy Future for the Pacific Northwest: An Alternative Scenario* (Palo Alto, Ca.: Natural Resources Defense Council, Inc., January 1977).

Chapter I. *Generation at the Turning Point*

1. *Reflections on the World Today* (New York, 1948), quoted in Daniel Bell, *The Coming of Post-Industrial Society* (New York: Basic Books, 1973), p. 345.

2. See Elting E. Morison, *From Know-How to Nowhere* (New York: Basic Books, 1974), p. 120.

3. "Roll on, Columbia," (New York: Ludlow Music, 1941). Quoted in Vera Springer, *Power and the Pacific Northwest* (Portland: Bonneville Power Administration, n.d. [1976]), epigraph.

4. Don Hodel, "The Prophets of Shortage," speech delivered to the City Club of Portland, July 11, 1975 (Portland: Bonneville Power Administration, 1975), p. 4.

5. Ibid.

6. Bonneville Power Administration, *The Role of the Bonneville Power Administration in the Pacific Northwest Power Supply System, Including Its Participation in the Hydro-Thermal Power Program,* draft environmental impact statement, Department of the Interior DES 77-21, July 22, 1977, Appendix C, p. IV-15. Hereafter this document will be referred to as the "BPA Role EIS."

7. Bonneville Power Administration, *1978 Annual Report* (Portland: BPA, 1979), p. 37.

8. See Ernst B. Haas, *The Obsolescence of Regional Integration Theory,* Research Series No. 25, University of California, Institute of International Studies, Berkeley, 1975, pp. 50-55.

9. S. David Freeman, *Energy--The New Era* (New York: Vintage, 1974), p. 325.

10. Comptroller General of the United States, *Region at the Crossroads--The Pacific Northwest Searches for New Sources of Electric Energy* (Washington: U.S. General Accounting Office, August 10, 1978), p. 2.5.

11. Ibid., p. 2.6.

12. *Power Review 77*, p. 68.

13. Bonneville Power Administration, *1978 Annual Report,* pp. 37, 18.

14. BPA Role EIS, Appendix A, p. III-102.

15. Ibid., Appendix A, Chap. III, Part A surveys the state of understanding.

16. Ibid., Appendix A, p. III-20.

17. Bonneville Power Administration, *Proposed 1979 Wholesale Rate Increase*, draft environmental statement, Department of Energy DOE/EIS-0031-D (Portland: BPA, 1978), pp. B-5, B-8.

18. *Power Review 77*, p. 68.

19. Roger Beers and Terry R. Lash, *Choosing an Electrical Energy Future for the Pacific Northwest: An Alternative Scenario* (Palo Alto, Calif.: Natural Resources Defense Council, Inc., 1977), pp. 23-24.

20. Washington Public Power Supply System, confirmed in a telephone interview, July 1979.

21. Bonneville Power Administration, *1978 Annual Report*, p. 50, apportions $5.4 billion to commercial power.

22. Ibid., p. 37.

23. Charles J. Hitch, "Energy in Our Future," *Key Reporter* (Phi Beta Kappa, Summer 1978), p. 2ff.

24. U.S. Senate, *Pacific Northwest Electric Power Planning and Conservation Act*, Report No. 96-272, to accompany S. 885, 96th Cong., 1st Sess., July 30, 1979, pp. 63, 65; the capital costs are estimated as "value of regional reserves."

25. Portions of the research reported here may be found in Kai N. Lee, Marion E. Marts, Robert G. Walton, and Alison D. Birmingham, "Electric Power Development in the Pacific Northwest Region: Institutional Commitments and Alternatives, Phase 1," project completion report, Institute for Environmental Studies, University of Washington, September 1976. See also Alison D. Birmingham, "Regional Electric Energy Planning: A Case Study in the Politics of Scarce Resources," unpublished M.U.P. thesis, University of Washington, 1976; Robert G. Walton, "The Development of the Electric Power System in the Pacific Northwest, a Public Policy Perspective," unpublished M.S.E. thesis, University of Washington, 1976; and Donna L. Klemka, "Pacific Northwest Electrical Energy Planning: Problems of Institutional Redesign," unpublished M.A. thesis, University of Washington, 1978.

Chapter II. *The Hydroelectric Heritage*

1. Quoted in Alexander Bickel, *The Least Dangerous Branch* (Indianapolis: Bobbs-Merrill, 1962), p. 16.

2. Compare, for example, John V. Krutilla, *The Columbia River Treaty* with Larrett Higgins, "Electricity and Canadian Policy," *The Energy Question*, Vol. 2, pp. 171-88. For a comprehensive account from the Canadian point of view, see Neil A. Swanson, *Conflict Over The Columbia*.

3. Higgins, "Electricity," p. 184.

4. The most exhaustive compilation of factual and statistical information on electric power in the Pacific Northwest is the BPA Role EIS.

5. Bonneville Power Administration, *1978 Annual Report*, p. 43.

6. *Power Review 77*, pp. 37 (hydro), 41 (thermal).

7. Ibid., p. 92.

8. BPA, *1978 Annual Report*, p. 34.

9. Stephen Breyer and Paul W. MacAvoy, "The Federal Power Commission and the Coordination Problem in the Electrical Power Industry," *Southern California Law Review*, 46 (1973) observe at p. 664 that:

   > The actual production of electricity accounts for about 53 percent of its cost, transmission for 12 percent, and distribution for about 35 percent. Several hundred firms owned by private investors provide approximately three-quarters of the nation's electricity supply. The remainder is produced by local, state or federal governments. Nearly all private firms are vertically integrated.

   The Pacific Northwest industry structure diverges markedly from the national norm. See also Theodore J. Nagel, "Operating a Major Electric Utility Today," *Science* 201 (1978), pp. 985-93.

10. George Tupper and Tom Noguchi, interview by DLK, Seattle, May 23, 1978. Tupper and Noguchi, of the Seattle area office of BPA, estimated production costs at the plant ("busbar") as follows:

Federal hydro (averaged)	1.2 mills/kwh	
Hanford (nuclear, online 1966)	4.5	
Centralia (coal, online 1975)	6.	
Trojan (nuclear, online 1975)	12.	(public share only)
Colstrip (coal, online 1975 and 1976)	12-14 mills/kwh	

11. BPA, *1976 Annual Report,* p. 38, note 7.

12. BPA, *About BPA,* p. 8.

13. American Public Power Association, "Tacoma's Low Power Costs Cited in Survey by TV Station," *Public Power Weekly.* January 16, 1978.

14. BPA, *1976 Annual Report,* p. 9.

15. Ibid., pp. 26-27.

16. For historical perspective on vertical integration, see Martin T. Farris and Roy J. Sampson *Public Utilities: Regulation, Management, and Ownership,* pp. 138-39. Compare the critique of vertical integration among IOUs in Edward Berlin, Charles J. Cicchetti, and William J. Gillen, *Perspective on Power,* Chap. 6.

17. William G. Hulbert, Jr., in Senate Hearings, p. 132.

18. BPA Role EIS, Appendix C, Attachment B, p. B-18.

19. Roy F. Bessey, *Pacific Northwest Regional Planning--A Review,* pp. 61-63.

20. Higgins, "Electricity and Canadian Policy," p. 185.

21. BPA Role EIS, Appendix B, p. I-1.

22. Breyer and MacAvoy, "The Federal Power Commission...," p. 689. The difficulties of sharing the costs and benefits of interconnection under existing regulatory arrangements are the focus of Breyer and MacAvoy's analysis.

In the Public Utility Regulatory Policy Act of 1978, one of the components of the National Energy Act, the Federal Energy Regulatory Commission (FERC) was authorized to order the physical interconnection of transmission grids, subject to formal procedures meant to assure that these interconnections would be in the public interest.

23. For a discussion of utility sensitivities, see Robert C. Walton, "The Development of the Electrical Power System in the Pacific Northwest," Chap. 3.

24. BPA Role EIS, Appendix B, p. II-1.

25. BPA, *1978 Annual Report,* p. 37.

26. BPA Role EIS, Appendix C, pp. IV-129 to IV-133; the employment figures reported are for 1973.

27. Ibid., IV-288 to IV-289.

28. Earl Gjelde, interview by DLK, November 7, 1977. Gjelde is assistant power manager of BPA.

29. This description follows Hussein Hassoun, *Industrial Energy Consumption Studies, Part I, The Aluminum Industry,* Appendix III.

The IF-1 schedule is administered through interim contracts promulgated in 1975. These interim contracts may be terminated on 30 days' notice by either BPA or the industrial customer. These agreements were established as part of a rate modification undertaken by Bonneville, as financial and legal pressures beset the agency. Although the DSIs agreed to trade in their existing power contracts and to switch to the new IF-1 rate schedule, new long-term industrial contracts could not be signed, pending completion of the BPA Role EIS. The environmental statement has not yet been issued in final form.

30. BPA Role EIS, Appendix C, pp. IV-134 to IV-142.

31. For a defense of the DSI role, see Harry V. Helton, Senate Hearings, pp. 151-66.

See also BPA Branch of Rates, "Bonneville's Consideration of Ratemaking Standards Under the Public Utility Regulatory Policy Act (PURPA)," discussion paper, June 1970, p. 11.

32. BPA Role EIS, Appendix C, p. IV-209, reports this price for January 1976; wholesale aluminum prices fluctuate substantially, however, in response to market demand. It is noteworthy that, during the 1974-75 economic recession, Pacific Northwest aluminum plants were among the last to be shut down as demand dropped, and were among the first re-started as the economy picked up. This provides unambiguous evidence that the region's plants offer economic advantages to their owners. See ibid., p. IV-205.

33. Ibid., p. IV-203. Year-long average costs for DSI power have been trending upward; until the early 1970s the averages were in the range of 2-2.5 mills/kwh.

34. Ernst and Ernst, *The Energy-Economy Relationship* (Washington, D.C.: report to BPA, 1976), discussed in BPA Role EIS, Appendix C, pp. IV-200 to IV-212.

35. Ibid., p. IV-210. It is important to keep in mind that the 9-12 mill/kwh rate is based on the assumption that the power would be supplied on a *firm* basis, rather than as interruptible power. Aluminum plants outside the Pacific Northwest are mostly supplied with firm power. The economic equivalent of firm is provided (in principle, at least) to Northwest firms through the BPA IF-1 rate schedule.

36. William B. Beyers, University of Washington, communication to DLK, March 1978.

37. Helton, Senate Hearings, p. 10.

38. Bonneville Power Administration, Branch of Rates, "Analysis of Rate Directives, Including Preference Customer Rate Limit," draft paper, February 1979, p. 13. A final version of this analysis has been printed as Appendix B of the Senate report on the regional power bill, U.S. Senate, "Pacific Northwest Electric Power Planning and Conservation Act," Report No. 96-272, 96th Congr., 1st Sess., 1979, p. 66.

39. BPA Role EIS, Appendix C, pp. IV-136 to IV-138.

40. The public power movement in the Northwest is discussed in Vera Edinger Claussen, *Public Power--Rags to Riches*.

41. Indeed, one of the landmarks of public power case law is *Puget Sound Power and Light Co.* v. *Seattle*, 291 U.S. 619 (1934) which established the principle that a municipally owned utility need not be operated at a profit. Cited and discussed in Farris and Sampson, *Public Utilities*, p. 266 and fn. 10.

42. For a detailed study of the early period, see Charles McKinley, *Uncle Sam in the Pacific Northwest*, especially Chapters 1-5. Bessey, *Pacific Northwest Regional Planning*, provides additional, though more fragmentary information up to 1963. A technically oriented history is found in Gordon D. Friedlander, "The Story of Bonneville Power: 1937-1968-1987," *IEEE Spectrum* (November 1968), pp. 78-94 and (December 1968), pp. 61-72.

43. McKinley, *Uncle Sam*, pp. 206-24; cf. the peaceful relations with the Army Corps, the other agency responsible for operating dams in the Federal Columbia River Power System, pp. 200-206.

44. See, *e.g.*, Grant McConnell, *Private Power and American Democracy* pp. 18-21; and Farris and Sampson, *Public Utilities*, Chapter 9, and pp. 270-74.

    The broader question of government competition is discussed in detail in Richard Hellman, *Government Competition in the Electric Utility Industry*, Chapter 2.

45. The reason government competition is needed at all is the "natural monopoly" character of electricity supply. Electric utilities are usually monopolies: only one seller is available to serve any particular customer, because to do otherwise would necessitate large duplicative costs for distribution systems, meters, and the like. In order to protect against the economic abuses of monopoly, government licenses utilities, and regulates the prices they may charge. In the absence of marketplace competition, however, the fair costs associated with the production and distribution of electric power are difficult to determine --unless there is an example provided, for example, by public production of power. See Hellman, *Competition in the Utility Industry*, pp. 60-62.

46. Ibid., Chapter 3, and pp. 22-23.

47. Ibid., p. 228; Farris and Sampson, *Public Utilities*, p. 293.

48. McKinley, *Uncle Sam*, pp. 9-11.

49. 16 U.S. Code Chapter 12B: Act of August 20, 1937, Ch. 720, 50 Stat. 731, as amended. The four policies discussed in the text are found in 16 U.S.C. 832a(b), 832c(a) respecting "domestic and rural" and "public" preferences, and 832d(a), respectively.

50. The Raker Act of 1913 granted San Francisco the right to build a dam on national park land in the Sierra Nevada Mountains, but restricted sale of the water or energy from the project to "a municipality or a municipal water district or irrigation district." In 1920 Congress created the Federal Power Commission and directed it to "give preference to applications...by states and municipalities" in granting permits and licenses. The 1928 Boulder Canyon Project Act, which authorized Hoover Dam, reaffirmed "the policy expressed in the Federal Water Power Act which created the Federal Power Commission as to conflicting applications," though some exceptions were spelled out. Public preference, sometimes extended to encompass federal loans for power development, was repeatedly embraced by Congress, in the Rural Electrification Act of 1936, the Fort Peck Project Act of 1938, the Reclamation Project Act of 1939, and the Flood Control Act of 1944.

51. For a fuller discussion of preference and its legal context, consult *Water Resources Law*, Volume 3 of the Report of the President's Water Resources Policy Commission, pp. 281-84, 298-99; Edward N. Fadeley, "Preference Rights of Publics and Cooperatives," and Bessey, *Pacific Northwest Regional Planning*, passim.

52. McKinley, *Uncle Sam*, pp. 157-70.

53. See, however, Harvey R. Doerksen *Columbia River Interstate Compact, Politics of Negotiation*, pp. 127-37. Doerksen concludes that the leading reason the compact--a water-resource development agreement--foundered in 1962 was that "the Compact became embroiled in an ideological conflict related to power marketing" (p. 181).

54. McKinley, *Uncle Sam*, p. 182.

55. Ibid., pp. 178-96.

56. Quoted in Owen, *TVA*, p. 97.

57. See Bessey, *Pacific Northwest Regional Planning*, pp. 85-89.

58. Krutilla, *op. cit.*, pp. 12, 26, and Chapter 3.

59. This account follows Walton, *op. cit.*, pp. 38-60.

60. Krutilla, *op. cit.*, pp. 43-44. The critical period provides a temporal base line for gauging the reliable power production capability of a hydroelectric system. It is defined as the maximum expected length of time which will elapse between refilling of the reservoirs in the system. The critical period begins when the reservoirs are full, and continues through the worst expected drought, through to its conclusion, ending when the reservoirs are once again full. A hydro system with a short critical period, such as the Columbia before the treaty, operates on an annual run of the river basis: productive assets--either spilled water or idle turbines--are left unused for part of each critical-period drought. (In years of noramal rainfall, there may be enough water to operate turbines throughout the year. But this cannot be counted on.) The firm power that can be generated throughout the critical period is defined as the reliable output of the system. The lengthening of the Columbia hydro system critical period by a factor of six increased the reliable capacity of the system, though not proportionally, as Krutilla explains.

61. Ibid., pp. 121-22.

62. See Marguerite Owen, *The Tennessee Valley Authority*, Chap. 2. For a broader perspective on TVA's institutional history, see ibid., passim, for a chronology, but compare Philip Selznick, *TVA and the Grass Roots* and James Branscombe, "Destroy to Save?" *Environment*, pp. 6-11, for critical appraisals. Deborah Shapley, "TVA Today: Former Reformers in an Era of Expensive Electricity," *Science* 814-18 provides interesting parallels to the problems encountered in the Pacific Northwest.

Chapter III. *Hydro-Thermal--The Second Generation*

1. Elting E. Morison, *From Know-How to Nowhere* (New York: Basic Books, 1974), p. 143.

2. This account follows Robert G. Walton, "The Development of the Electrical Power System in the Pacific Northwest, A Public Policy Perspective," (M.S.E. thesis, University of Washington, 1976), pp. 28-38.

3. Cleveland Anschell, "Government City, Richland, Washington," *Urban Trends* (Rainier National Bank newsletter) 1 (1978), pp. 6-8, reviews the intimate association between nuclear energy and the rapid growth of the Tri-Cities region.

4. Bonnie Baack Pendergrass, "Public Power, Politics, and Technology in the Eisenhower and Kennedy Years: The Hanford Dual-Purpose Reactor Controversy, 1956-1962," (Ph.D. diss., University of Washington, 1974), p. 44.

5. Ibid., p. 123.

6. Vera Edinger Claussen, *Public Power--Rags to Riches* (Seattle: Washington Public Utility Districts Association, n.d.), p. 23.

7. This account follows Dan Seligman, "The Washington Public Power Supply System" (unpublished paper, University of Washington, Institute for Environmental Studies, December 1978, Appendices A and B).

8. The largest utility members, Seattle, Tacoma, and Snohomish County, are traditionally represented on the executive committee responsible for day to day oversight of WPPSS management.

   It is worth pointing out that WPPSS, as a major sponsor of thermal power in the region, exercises important public responsibilities. Unlike an investor-owned utility, however, the Supply System is not subject to economic regulation by a public commission. Such oversight as is exercised comes from two sources: Bonneville, which is responsible, through the financial arrangements discussed below, for regional guarantees of power supply covering three WPPSS nuclear plants; and the board of directors of the Supply System. This board, heavily weighted toward small, rural PUDs, can hardly be said to be representative of the broad interests of the State of Washington, despite the heavier representation accorded the major urban centers by membership on the executive committee. This uneven representation is in accord with federal law, which exempts administrative agencies composed of representatives from elected bodies from the one man-one vote rule governing state legislatures. (See *Sailors* v. *Board of Education*, 387 U.S. 105; we are grateful to Professor Race Davies for this citation.)

9. This was pointed out to us by Eric Redman, private communication, July 1978; see also BPA Role EIS, Appendix A, p. IV-107.

10. Donald A. Pugnetti, "Jackson Offer on Hanford Steam Plant Is Studied," *Tri-City Herald* (July 12, 1962), p. 1.

11. U.S. Army Corps of Engineers, *Review Report on Columbia River and Tributaries*, House Document 403, 87th Cong. 2nd Sess., 1958. This report revised and updated the 1948 version of the same title, H. Doc. 531, 81st Cong., 2nd Sess., 1948. An earlier precedent, though not so systematically argued, may be found in Holland H. Houston, *The Place of Steam-Electric Generating Stations in the Orderly Program of Electric Power Development in the Pacific Northwest*, Bulletin 38, Department of Conservation and Development and Public Service Commission, State of Washington (Olympia, 1950), p. 5; Houston, a proponent of thermal power, projected a need for thermal generation by 1965.

The hydro-thermal notion was clearly in the air. In 1956, the Washington State Power Commission observed, "Use of steam generation in conjunction with hydro plants makes it possible in some cases to increase the firm output of a power system at a lesser cost than by construction of new hydroelectric projects or by the use of steam plants alone." Commission report of May 7, 1956 (Seattle and Olympia), p. 12. In November 1956 the commission issued a preliminary report, *Steam-Electric Generating Facilities*, embodying a study done by the engineering firm of Ford, Bacon & Davis; this report included an estimate of the coal available to support a 311 megawatt thermal plant at Lake Cle Elum, Washington.

12. See Paul L. Joskow and Martin L. Baughman, "The Future of the U.S. Nuclear Energy Industry," *Bell J. Econ.* 7 (1976) pp. 3-32, esp. pp. 8-12. The applicability of these ideas to the fragmented utility industry of the Northwest is argued in Hussein Hassoun, Joseph Nadal, and Richard Timm, "Testimony in the Matter of the Application of Portland General Electric Company for a Site Certificate for the Pebble Springs Nuclear Power Plant," before the Oregon Energy Facility Siting Council (Salem: Oregon Department of Energy, April 28, 1976, processed).

13. The percentages are derived from Philip G. Hill, *Power Generation* (Cambridge: MIT Press, 1977) p. 220:

technology / sponsor	coal (load factor = 80%)	nuclear (load factor = 60%)
public agency (interest rate = 10%)	51	78
IOU (interest rate = 16%)	62	85

Percentage of Generation Costs Due to Capital Charges

These figures are estimated for 1000 megawatt plants coming on-line in 1984, with the costs of power estimated to lie between 22 mills/kwh (for a nuclear plant built by a public agency) and 32 mills/kwh (for a nuclear plant built by an investor-owned utility).

14. Other regions of the nation with extensive hydro resources, such as the Tennessee Valley, have also used this flexibility of hydroelectric generation in the development of thermal power generation.

15. Bonneville Power Administration, *A Ten Year Hydro-Thermal Power Program for the Pacific Northwest* (Portland, 1969), pp. 31, 19. No load factor assumption for thermal plants is stated explicitly, although the ratio of average to peak capacity in the BPA analysis for the late 1980s is about 85 percent. Ibid., p. 50.

16. Northwest Energy Policy Project, *Energy Futures Northwest* (Springfield, Va.: National Technical Information Service, 1978), p. 20; *Power Review* 76, p. 52.

17. Donald Paul Hodel, interview by KNL, Portland, February 1978, transcript (revised, March 1978) on file at the University of Washington.

18. Bernard Goldhammer, interview by MEM, Robert G. Walton, and Alison D. Birmingham, Portland, April 1976, transcript on file at the University of Washington.

19. Ibid.

20. Ibid.

21. Hodel, to KNL, February 1978.

22. Bernard Goldhammer remembered state governments as ineffectual onlookers in HTPP's founding: "We've got this problem in the states--the person who should be involved in the decision making is the governor, and he's got so many other things to do he can't be involved, really. So he assigns somebody else [to it] and we run into the various differences of the offices of the governor and the states' [bureaucracy], where there's complete disagreement." Goldhammer to MEM, Walton and Birmingham, April 1976.

23. Hodel to KNL, February 1978.

24. BPA Role EIS, Appendix C, p. II-46; the IF-1 rate schedule is discussed in Chap. 1.

25. Goldhammer to MEM, Walton and Birmingham, April 1976.

26. Don Hodel, private communication, April 1978.

27. The ability of public agencies to raise money for thermal plants may have been underestimated by all concerned, however: "We long thought they would hit a limit beyond which they couldn't go, but they haven't found that limit

yet." Hodel interview, KNL, February 1978. The credit guarantees described in the text have probably extended the limit considerably--assuming such a financing constraint is well defined in the first place.

28. See Richard Hellman, *Government Competition in the Electric Utility Industry* (New York: Praeger, 1972), pp. 48-51, for a comparison of private sector sensitivities to various governmental undertakings in power supply.

29. Hodel to KNL, February 1978. See also the account of the Dixon-Yates controversy in the Tennessee Valley, Marguerite Owen, *The Tennessee Valley Authority* (New York: Praeger, 1973), pp. 103-18.

30. A nuclear plant with a nominal 1000 megawatt output will not operate at full rated capacity year round. There will be shutdowns for refueling, maintenance, and repairs, as well as operation at less than full output for economic or technological reasons. A typical "load factor," measuring the average power output, is 60 percent for nuclear plants. This would place the output of a 1000 megawatt plant at 600 megawatts averaged over the year.

31. The "requirements contract" with preference customers should be contrasted with contractual arrangements with DSIs to provide "industrial firm" power. Industrial firm implies only limited responsibility by BPA to provide power. Instead, portions of the DSIs' power allotment are to be used to cover others' shortfalls, with industrial customers receiving credits to compensate them economically for power undelivered.

32. This necessity for congressional approval by the appropriations committees was pointed out to us by Don Hodel, private communication, April 1978.

33. Goldhammer to MEM, Walton and Birmingham, April 1976.

34. Anonymous interview by KNL, June 1979.

35. Ibid. After net billing had been approved by Congress, GAO and BPA fought a bureaucratic skirmish about Bonneville's responsibility to disclose the financial implications of net billing; BPA ultimately consented to include a footnote in its annual statement. By 1979, this footnote had been augmented by the explicit statement that net billing credits have first priority among Bonneville's financial obligations, ahead of debt service on the

federal hydro system. BPA, *1978 Annual Report,* pp. 45 (note 1), pp. 47-48 (note 5); cf. BPA, *1976 Annual Report,* pp. 37-38.

36. Alfred E. Kahn, *The Economics of Regulation* (New York: Wiley, 1970). Chap. 1, 2, presents the standard argument. Note that marginal cost-pricing presents particular difficulties in capital-intensive industries. The basic problem is that the costs of production at any given time reflect an average over capital investment costs over a long period of time in the past. When costs are falling--as they have traditionally done in electric power generation --the marginal cost is falling too, since today's technology can produce a kilowatt-hour more cheaply than yesterday's. Generally, yesterday's facilities have not yet been paid for, however, and marginal cost pricing would bankrupt the utility by providing insufficient revenue to pay back investment costs. When--as has been the case recently--production costs are increasing, marginal cost prices are far above the actual, average cost of producing for today, since today's technology now costs more than yesterday's--but yesterday's facilities are still producing. Now, large revenues are produced by marginal cost prices.

    The economist's prescription in the latter case is to charge marginal costs so as to signal the costs of growth to the consumer. At the same time, however, government should impose an excess-profits tax, to divert the large returns away from utilities, redirecting them to broader social purposes. This recipe has proved unappealing to politicians, as the difficulties of the Carter Administration's petroleum pricing proposal, which employs a marginal cost approach, illustrates.

37. For an overview, see Hill, *Power Generation,* Chap. 7.

38. This account follows Seligman, "Washington Public Power Supply System," pp. 4, 5.

39. Ibid., p. 12.

40. Bonneville Project Act, 16 U.S. Code 832d(a).

41. Branch of Customer Service, BPA, "Net Billing Analysis," unpublished memorandum, January 3, 1973. We are grateful to George Reich for providing this memorandum.

42. Kell, Alterman and Runstein, *Laws Needed to Expedite Electric Generation Schedules for the Pacific Northwest: Financing, Environmental Laws Siting and Licensing Procedures* (Portland, 1976), Vol. 1, pp. 145-46.

43. Joel Merkel, interview by KNL, Washington, June 1979.

44. John Kenneth Galbraith, *The New Industrial State,* rev. ed. (New York: New American Library, 1971), Chap. 2.

45. Ibid., Chap. 6-8.

46. A recent discussion paper written by BPA staff suggests the durability of the hydro-thermal concept. Their analysis of rates to be charged for hydro-thermal power still ignores the environmental costs associated with load-following operation of the hydro dams--a problem described in Chapter IV. Moreover, long-run incremental-cost pricing, the approach favored by economic efficiency, is still dismissed as lacking in support. Such support is vital, of course, if coordinated planning (an activity that is political as much as it is economic) is to be sustained. See BPA Branch of Rates, "Bonneville Power Administration's Consideration of Ratemaking Standards Under the Public Utility Regulatory Act (PURPA)," discussion paper, June 1979.

47. Douglas Woodfill, "Pacific Northwest Power Entities, An Expanded Glossary," unpublished, Seattle City Light, December 1975, p. 11.

48. See, generally, Pacific Northwest River Basins Commission, Power Planning Committee, *Power Planning Organizations in the Pacific Northwest,* July 1978.

49. For additional background see Donna Lee Klemka, "Pacific Northwest Electrical Energy Planning," (M.A. thesis, University of Washington, 1978), Appendix B.

50. Compare James D. Thompson, *Organizations in Action* (New York: McGraw-Hill, 1967), p. 42, for a sociological discussion indicating that growth is an imperative generally.

51. The West Group Forecast is published annually in two parts, a summary document--the "Black Book"--issued in March, and the full 20-year forecast--the "Blue Book"--in April. Eleven public agencies and four IOUs submit forecasts prepared by their own staffs to the PNUCC; Bonneville helps to develop forecasts for the other West Group public

utilities, listed as "other public agencies" in PNUCC tables and documents.

52. Oregon Department of Energy, *Oregon's Energy Future--First Annual Report* (Salem: Oregon DOE, 1977), esp. p. 35; Department of Energy, *Oregon's Energy Future--Second Annual Report* (Salem: Oregon DOE, 1978), esp, p. 71.

53. The analysis in this series with the broadest regional planning significance is the questioning of the critical water year planning assumption in Hassoun, Nadal, and Timm, "Testimony on Pebble Springs Nuclear Power Plant," April 1978.

54. In his attack on environmentalists, whom he accused of being obstructionists and "prophets of shortage," BPA administrator Hodel said of energy conservation, "these kinds of vital, wrenching decisions should not be left to the utilities to make." "The Prophets of Shortage (Part 2)," remarks to the Rotary Club of Seattle, August 27, 1975 (Portland: BPA, 1975), p. 8.

55. Washington Public Power Supply System, *1980 Budget* (Richland, Wash.: WPSS, 1977).

56. WPPSS, *1977 Annual Report*, p. 23.

57. Seligman, "Washington Public Power Supply System," pp, 11-12 and Appendix I.

58. John Kenneth Galbraith, *Economics and the Public Purpose* (New York: New American Library, 1973), p. 39.

59. Norman A. Stoll, counsel to the Public Power Council, expressed his concern "as to whether all of the preference agency policy-making officials sufficiently understand the program" proposed in 1977 by the Pacific Northwest Utilities Conference Committee that would revive the Hydro-Thermal Program. "Pros and Cons of Proposed Regional Power Supply Program as Developed to Date," memorandum for PPC Executive Committee, April 22, 1977, unpublished, p. 26. Chapter V will discuss in more detail the problems of representing the interests of public power, especially the smaller utilities influential in PPC, in the formulation of the PNUCC legislative proposal.

60. Galbraith, *The New Industrial State*, pp. 49-50

61. Ibid., Chap. 25

Chapter IV. *The Broadening Agenda*

1. Quoted in Arthur Magida, "Movement Undaunted by Economic, Energy Crises," *National Journal* 8 (1976), p. 61.

2. Don Hodel, "The Prophets of Shortage (Part 2)," speech delivered to the Rotary Club of Seattle, August 27, 1975 (Portland: BPA, 1975), p. 2.

   William H. Rodgers, Jr., "Siting Power Plants in Washington State," *Wash. Law Rev.* 47 (1971) 10n, cites Federal Power Commission figures showing that only 6 percent of the delays in power plant construction are due to environmental intervention.

3. Hodel, "Prophets of Shortage," p. 3.

4. BPA Role EIS, Appendix A, pp. III-102 and III-103.

5. Ibid., p. III-107.

6. Ibid., p. III-104. (On squawfish, see pp. III-116, 118.)

7. Ibid., pp. III-105 to 107.

8. Among the major uncertainties are the operating schedules for upstream storage reservoirs and the proportion of peaking to baseload hydropower demanded by utilities.

9. Ironically, construction of the Third Powerhouse at Grand Coulee has increased the physical ability to provide flood control at exactly the time when the economic costs of flood control have risen. The powerhouse permits much faster drawdown of Lake Roosevelt, by enlarging the outlet capacity substantially. Limited outlet capacity has been a problem in the past because of the large volumes of flood flows.

10. The loads of the direct-service industrial customers, which do not vary during the day or year, will remain constant in the future; this will reduce the base load as a proportion of the overall average load. If improved demand management is instituted--probably through interchanges among utility systems with different patterns of demand fluctuation--the contribution to the peakload from load growth may also be held down. See BPA Role EIS, Appendix A, pp. IV-77 to 96.

11. Ibid., Figure III-10, p. III-20. See also p. III-16 for a discussion of the conservative character of the estimates. U.S. Heritage Conservation and Recreation Service, *Impacts on Recreation from Hydroelectric Power Peaking of the Mid-Columbia River, Washington* (Seattle: Northwest Regional Office, July 1978) Appendix III reports that variations in water level of twenty-six feet may result from maximum development of Grand Coulee dam hydropower peaking by 1995. The twenty-six foot estimate is for a site seven miles downstream from the dam's tailwaters, where the magnitude of the elevation change is already diminished.

12. Ibid., pp. III-7, III-103.

13. Ibid., pp. III-162, III-166.

14. Ibid., p. 166.

15. Ibid., p. 169.

16. Ibid.

17. PNUCC Task Force 8 (Thermal Plant Operating Characteristics), Report to the Technical Coordinating Group, September 20, 1977. Quotations for "Inability of the System to Meet Forecasted Loads," p. 1.

18. The cost estimate is by Norman K. Whittlesey et al., *Benefits and Costs of Irrigation Development in Washington,* prepared for the House of Representatives and the Legislative Transportation Committee, Washington State Legislature, October 1976, Vol. I, p. 20. These estimates assume an energy replacement value of 20 mills/kwh. The acreage figure for 1975 is in BPA Role EIS, Appendix A, pp. III-143-44. In fiscal 1978, sales of electricity by the Federal Columbia River Power System brought revenue of $267 million. (BPA: *1978 Annual Report,* p. 37).

19. Bonneville Power Administration, *1977 Annual Report* (Portland: BPA, 1978), p. 21.

20. Most hydroelectric systems are, in utility parlance, "energy limited," while thermal systems are "peak limited." This means that, as the level of average demand rises, a hydro system will generally run out of average energy before its ability to deliver peak power is exhausted. A thermal system, as it grows, experiences the reverse order of shortage. This historical experience describes the evolution of the Pacific Northwest power supply system,

which is now under increasing pressure to increase its average energy supply. Assuming that is successful, however, there will be gradually increasing demand for peak power--and thus the program to increase the peaking capability of the federal dams. See BPA Role EIS, Appendix A, p. II-7.

21. U.S. Army, Corps of Engineers, *Pumped-Storage Inventory of the Pacific Northwest,* Report No. 26, North Pacific Division, Portland, Oregon, January 1976, lists 530 sites, "each with a capability of furnishing at least 1,000 megawatts of capacity on a weekly operating cycle." In addition, four seasonal pumped-storage sites are included.

22. William M. Ross and Marion E. Marts, "The High Ross Dam Project," *Canadian Geographer* 19 (1975), pp. 221-34.

23. Polly Dyer, interview by KNL, Seattle, April 1978.

24. For an account of the regulatory difficulties facing national coal policy, see Richard L. Gordon, "The Hobbling of Coal," *Science* 200 (1978), pp. 153-58.

25. The Centralia plant emits enough sulfur dioxide to increase the ambient concentration by 50 to 100 micrograms per cubic meter. The primary air pollution standard, which sets the limit deemed safe for the protection of human health, is 260 micrograms per cubic centimeter. Existing sources of sulfate pollution already cause peak concentrations of 150; the addition of the Centralia project thus presses upon the allowed carrying capacity of the airshed. For this reason, the construction of a third unit at Centralia has been indefinitely deferred. Lee Bamesberger, Department of Chemical Engineering, Washington State University, telephone interview by KNL, July 1976.

26. Tom McGrath, Puget Sound Power and Light, telephone interview by KNL, October 1978. See also Luther J. Carter, "SO_2 Emissions Pose Growth Issue," *Science* 202 (1978), p.30.

27. Northern Plains Resource Council, "Future Northwest Power Needs Pose Serious Threat to Plains States," *Washington Resource Report* (April 1978), pp. 4-5.

28. See, *e.g.*, Thomas L. Judge, in Senate Hearings, pp. 524-25.

29. See, *e.g.*, John S. Gilmore, "Boom Towns May Hinder Energy Resource Development," *Science* 191 (1976), pp. 535-40.

30. John Harte and Mohamed El-Gasseir, "Energy and Water," *Science* 199 (1978), P. 632.

31. For a middle-of-the-road overview, see Nuclear Energy Policy Study Group, *Nuclear Power Issues and Choices* (Cambridge: Ballinger, 1977), Chaps. 3, 7-10. The problems of the industry are surveyed in Juan Cameron, "Where *Does* the President Stand on Nuclear Energy?" *Fortune* (March 1978), 99ff. See also Christoph Hohenemser, Roger Kasperson, and Robert Kates, "The Distrust of Nuclear Power," *Science* 196 (1977), pp. 25-34.

32. John Walsh, "Fuel Reprocessing Still the Focus of U.S. Nonproliferation Policy," *Science* 201 (1978), pp. 692-97. For a detailed account, see Gene I. Rochlin, *Plutonium, Power, and Politics* (Berkeley: University of California Press, 1979).

33. U.S. Energy Research and Development Administration, *Alternatives for Managing Wastes from Reactors and Post-Fission Operations in the LWR Fuel Cycle*, ERDA 76-43 (Springfield, Va.: National Technical Information Service, 1976). Interagency Review Group on Nuclear Waste Management, *Report to the President*, TID-29442, March 1979 (Springfield, Va.: NTIS, 1979) is somewhat more vague on timetables for resolution of the issue.

34. J. Gustave Speth, speech to the American Law Institute/American Bar Association Environmental Law Conference, September 29, 1977, Washington, D.C. Reported in *Environmental Reporter* "Current Developments" 8:23 (October 7, 1977), p. 855.

35. 547 F. 2nd 633 (1977) subsequently remanded to the Court of Appeals by the U.S. Supreme Court; see 11 ERC 1439 (1978).

36. A.B. No. 2822, Section 1, adding Public Resource Code Sec. 25524.2; discussed in "Fallout from the California Nuclear Initiative," *Environmental Law Reporter* 6 (1976), p. 10174, and in Luther J. Carter, "Nuclear Initiative," *Science* 192 (1976), pp. 975-77, 1317-19.

37. Peter Stoler, "The Irrational Fight Against Nuclear Power," *Time* (September 25, 1978), pp. 71-72.

For a recent review, see Dorothy Nelkin and Susan Fallows, "The Evolution of the Nuclear Debate: The Role of Public Participation," *Annual Review Energy* 3 (1978), pp. 275-312.

38. For a chronological account see John Walsh, "Opposition to Nuclear Power: Raising the Question at the Polls," *Science* 190 (1975), pp. 964-66; Philip M. Boffey and Nicholas Wade, "The Nuclear Debate: Clashes in Congress and California," ibid., 191 (1976), pp. 47-50; and Luther J. Carter, "Nuclear Initiatives: Two Sides Disagree on Meaning of Defeat," ibid., 194 (1976), pp. 811-12. Compare Edward Ricco, "Anti-Nuclear Sentiment Makes Itself Felt," *Nat. Res. J.* 19 (1979), pp. 377-79.

39. See Hussein Hassoun, Joseph Nadal, and Richard Timm, Testimony before the Oregon Energy Facility Siting Council, April 28, 1978) (Salem: Oregon Department of Energy, processed).

40. David Myhra, "One Nuke Gets a Warm Welcome," *Planning* (November 1975), pp. 13-18.

41. For a parallel to Leed, see Frank Graham, Jr., "The Outrageous Mr. Cherry and the Underachieving Nukes," *Audubon* (September 1977), pp. 50-67.

Also see William Tucker, "Environmentalism and the Leisure Class," *Harpers* (December 1977), pp. 49-80, an "expose" of environmental litigation in the Storm King pumped storage plant controversy outside New York City. On the merits of the case, see Luther J. Carter, "Con Edison: Endless Storm King Dispute Adds to Its Troubles," *Science* 184 (1974), pp. 1353-58.

42. David Dinsmore Comey, "Will Idle Capacity Kill Nuclear Power," *Bulletin of the Atomic Scientists* (November 1974), pp. 23-28.

43. The NRC licensing board at Skagit appears to be especially sensitive to seismic hazard in the wake of the discovery of a fault near the Diablo Canyon nuclear project north of Santa Barbara, California. The fault may force rescinding of the NRC operating license for the first nuclear plant at Diablo Canyon and halt construction on the second.

44. This discussion is based on Dan Seligman, "The Washington Public Power Supply System" unpublished paper, University of Washington, Institute for Environmental Studies, December 1978.

For a useful supplement on the organizational problems attending the complexity of nuclear technology, see Deborah Shapley, "Nuclear Power Plants: Why Do Some Work Better Than Others?" *Science* 195 (1977), pp. 1311-13.

45. The investment allocated to commercial power production in the Federal Columbia River Power System is $5.4 billion. Bonneville Power Administration, *1978 Annual Report*, p. 50.

46. For a discussion of the causes, see Philip G. Hill, *Power Generation* (Cambridge: MIT Press, 1977), Chap. 7.

47. The decision processes of the technostructure discussed in Chapter III significantly enhance the advantages of dealing with innovative, but tractable, large-scale technologies. See John Kenneth Galbraith, *The New Industrial State* (New York: New American Library, 1971), Chap. 2, 6, and pp. 177-78.

48. U.S. Nuclear Regulatory Commission, *Nuclear Energy Center Site Survey*, NUREG 76-0001 (Springfield, Va.: National Technical Information Survice, 1976).

C.C. Burwell, M.J. Ohanian, and A.M. Weinberg, "A Siting Policy for an Acceptable Nuclear Future," *Science* 204 (1979), pp. 1043-51, have analyzed the technical feasibility of expanding sites already containing nuclear plants. As controversy over the Pebble Springs, Oregon, site continues Portland General Electric, the project's principal sponsor, has publicly announced its interest in relocating these two nuclear plants at Hanford.

49. Marion E. Marts, "When Can California Join the Union?" *Association of Pacific Coast Geographers Yearbook* 23 (1961), pp. 7-12, takes issue with the traditional concern that the arid Southwest might seek to capture the abundant water in the Pacific Northwest. See also Marts' article in *Argus* (November 21, 1975). A ten-year moratorium on studying inter-basin water transfers, which ended in 1978, was given a ten-year extension by Senator Henry Jackson in the turbulent closing days of the 95th Congress. But the possibility that either water or power might have to be transferred south remains sufficiently

plausible to feed regionalist paranoia.

50. The interregional exchange agreement provides for Northwest peaking capacity to be made available to Southwest utilities; in return, Southwest utilities transmit energy to the Northwest. The exchange operates on a seasonal basis, making use of the historical differences in the profile of annual demand. Northwest loads are heaviest in wintertime, when electric heating causes high demand; this is also the season when river flow for hydro generation is at its yearly low. California and Southwest loads peak in summer, owing to demand caused by air conditioning; their utilities are based heavily on thermal generation, although California still relies substantially on hydropower generated in the Sierra Nevada mountains. Over the year, then, the Northwest experiences an annual drain on its energy capability in the winter, when loads are high and generation is reduced; the Southwest experiences high peaking loads in the summer, loads which are costly to meet using thermal generation, and much more cheaply provided by hydro. (The logic of this argument follows the analysis of the hydro-thermal concept in Chapter III.)

The seasonal exchange accordingly provides for Northwest provision of seasonal peaking capacity to the Southwest in the summer; in return, Southwest utilities provide energy in the winter, to help meet the Northwest load. The approximate ratio of exchange is one megawatt of capacity furnished by the Northwest, in return for 2500 megawatt-hours of energy from the Southwest. This results, on average, in a net gain in energy for the Pacific Northwest, equivalent to between 100 and 500 average megawatts of supply, according to George Reich, Bonneville Power Administration, conversation with KNL, November 1978.

The seasonal interregional exchange typically takes up one-fifth of the transmission capacity of the intertie. In addition, power surplus to the needs of the Northwest is transmitted southward; the amount of power sold as surplus in a given year depends upon the availability of water for secondary hydro generation.

Note also that the rapid growth of irrigation loads to serve the arid eastern regions of the Northwest may alter the seasonal variation in load, increasing demand on Columbia River hydro in summer, and eventually requiring modification of the interregional exchange.

51. William A. Brewer, conversations with MEM and KNL, 1974-75.

52. Roger M. Williams, "Oregon: The Fight for 'Survival'," *S/R World* (November 16, 1974), pp. 10-15.

53. Bardsley and Haslacher, Inc., "Electric Energy and the Environment," November-December 1975. We are indebted to Gordon C. Culp, attorney for the PNUCC, for making this study available to us.

54. Ibid., p. 11.

55. Robert Cameron Mitchell, "The Public Speaks Again: A New Environmental Survey," *Resources* No. 60 (September-November 1978), pp. 1-6; quotation from p. 1.

56. Pacific Northwest River Basins Commission, *Review of Power Planning in the Pacific Northwest, Calendar Year 1975* (Vancouver, Wash.: The Commission, 1976), pp. 20-21.

57. This account follows Jeffrey P. Foote, Alan S. Larsen, and Rodney S. Maddox, "Bonneville Power Administration: Northwest Power Broker," *Environmental Law* 6 (1976), pp. 831-57; and BPA Role EIS, Summary Report, pp. 1-2.

58. The tax status of the generating facilities under net billing was somewhat cloudy because of Bonneville's acquisition of power. In 1972 the U.S. Treasury issued regulations interpreting tax reform legislation enacted in 1968. The effect of this ruling was that BPA's ability to purchase output from publicly owned plants was restricted. Otherwise, the exemption from federal income tax of the revenue bonds used to finance the plants would be voided.

59. Seligman, "Washington Public Power Supply System," pp. 10-11.

60. BPA Role EIS, Appendix C, p. II-48. See Chapter V for a longer discussion of the reserve role of the DSIs.

61. Foote, Larson, and Maddox, "BPA: Northwest Power Broker," pp. 847-49.

62. *Port of Astoria* v. *Hodel*, 8 ERC 1156 (D. Ore. 1975).

63. *NRDC* v. *Hodel*, Civ. No. 75-344 (D. Ore. 1975).

64. As if to say that governmental troubles come in triplicate, BPA was the target of a third suit, *Weaver* v. *Hodel*, Div. No. 75-955 (D. Ore. 1975), filed by Congressman James Weaver, an environmentalist Democrat from Lane County, Oregon. Weaver attacked the IF-1 contracts on the ground that they failed to meet the test of "sound business practices" required of the Bonneville administrator by the Bonneville Project Act. *Weaver* v. *Hodel* was dismissed after administrator Hodel filed an affidavit freezing the terms of DSI contracts while the Role EIS was being prepared; this policy is still in effect.

65. Greg Hill, "The Politics of Energy 1990," unpublished paper, University of Washington, Institute for Environmental Studies, April 1978.

66. *Energy 1990 Consultants' Report* (Bellevue, Wash.: Northwest Environmental Technology Laboratories, 1976).

67. Hill, "Politics of Energy 1990."

68. Quoted in ibid., p. 49.

69. See Randy Revelle, in House Commerce Hearings, p. 367. The 1978-79 demand figure was reported by Gerald Garman, power manager of Seattle City Light, interview by KNL, July 1979.

70. Hill, "Politics of Energy 1990," p. 34.

71. Ibid., p. 36.

72. The emergence of a consensus among liberals and conservatives was similarly the basic political dynamic behind the institutionalizing of U.S. macroeconomic policy in the 1964 Kennedy Tax Cut. See Herbert A. Stein, *The Fiscal Revolution in America* (Chicago: University of Chicago Press, 1968), Chap. 16, 17.

73. Hill, "Politics of Energy 1990."

74. Ibid., p. 2.

75. Robert Wiebe, *The Search for Order, 1877-1920* (New York: Hill and Wang, 1967), esp. Chap. 5-8.

76. Edward C. Banfield identifies a process he calls "middle- and upper-classification":

> Half a century or more ago, the basis of city and state political power...was the machine. The bosses who ran it kept themselves in power by dispensing patronage...and therefore could pretty well disregard public opinion when it suited them to do so. Middle- and upper-classification rendered this system obsolete and brought into being one in which the politician, in order to compete successfully for office, has to combine offers of benefits to *classes* of voters (homeowners, taxpayers, and so on) with appeals to general ideas and conceptions of public interest. Whereas the old system had promised personal rewards, the new one promises social reforms.

The Unheavenly City Revisited (Boston: Little, Brown, 1974), p. 276.

For a discussion of public interest law, see Stuart A. Scheingold, *The Politics of Rights* (New Haven: Yale University Press, 1974), Chap. 12, and, more pessimistically, Tucker, "Environmentalism and the Leisure Class."

77. On technology as legislation, see Langdon Winner, *Autonomous Technology* (Cambridge: MIT Press, 1977), pp. 317-25.

78. Hill, "Politics of Energy 1990," p. 49.

79. Ibid.

80. See ibid., pp. 11-12, for a discussion of how political control of the consultants' assumptions became an issue.

The economist who prepared the econometric forecast, Donald Shakow, saw at the outset that City Light's assumption that surplus power from Seattle's share of WNP 4 and 5 could be sold at full cost was worth challenging. City Light superintendent Gordon Vickery insisted that this assumption be maintained, but when Shakow threatened to resign, Vickery backed down in the face of pressure from the COC. The compromise reached was that the consultant would do two forecasts, one assuming that SCL would be able to sell its surplus at cost, and one that assumed sales below cost.

Chapter V. *The Dance of Legislation*

HEADNOTE

Congressional bills. Three essentially different legislative proposals have been put forward in Congress since March 1977. In this chapter we shall refer to them by the following shorthand:

PNUCC bill. The Senate version, numbered S. 2080, was introduced September 9, 1977, by Henry Jackson, Democrat of Washington; its companion version in the House of Representatives was numbered H.R. 9020, and introduced by Lloyd Meeds, Democrat of Washington. Neither version advanced beyond congressional hearings.

Jackson bill. S. 3418 was introduced by Jackson in the second session of the Ninety-fifth Congress, on August 16, 1978; Meeds put forward a House version, numbered H.R. 13931. Neither bill advanced to committee consideration. On April 5, 1979, the bill was reintroduced by Jackson in the Ninety-sixth Congress under the number S. 885; in substantially amended form S. 885 passed the Senate on August 13, 1979. The companion House version, H.R. 3508, introduced by Al Ullman, Democrat of Oregon, awaited consideration in the fall of 1979, as this book went to press.

Weaver bill. Although introduced earliest, on March 31, 1977, H.R. 5862, the legislative proposal of Representative Jim Weaver, Democrat of Oregon, has received little serious attention. The Weaver bill has been repeatedly revised, its most recent version being H.R. 4159, introduced on May 21, 1979, to the Ninety-sixth Congress. Many ideas from Weaver's drafts have made their way into the regional power debate, and some survived in S. 885. The Weaver bill has not, however, received formal consideration beyond hearings by the House.

1. Woodrow Wilson, *Congressional Government*, p. 297.

2. Robert Gilpin, *U.S. Power and the Multinational Corporation*, p. 256.

3. Oregon State Legislative Assembly, S.B. 320, 1977 Regular Session. Governor Vic Atiyeh had not yet activated DRPA as of September 1979.

4. U.S. Congress, Senate, "A Bill to make the Federal Columbia River Power System available for maximum electric efficiency, for future essential power supply, to promote conservation, and for other purposes." The PNUCC bill (as we shall refer to it) would have been named the Pacific Northwest Electric Power Supply and Conservation Act (PNEPSCA), and it is often referred to by this acronym and by its Senate bill number.

5. U.S. Senate, "A BILL To assist the electric consumers of the Pacific Northwest...." This, the "Jackson bill," was introduced again on April 5, 1979, under the number S. 885, and its House version, introduced by Al Ullman, Democrat of Oregon, was numbered H.R. 3508. The quotation comes from Congressional Record 125, No. 44 (April 15, 1979).

6. E.g., Pacific Northwest Utilities Conference Committee, *Long Range Projection of Loads and Resources 1977-78 through 1996-97, West Group Area*. This forecast is commonly referred to as "The Blue Book."

7. Ibid., and GAO Report, p. 5.2 and Figure 5.1.

8. *NRDC Alternative Scenario*.

9. Ibid., pp. 28-29.

10. George Hinman et al., *Energy Projections for the Pacific Northwest*.

11. Oregon Department of Energy, *Oregon's Energy Future, Second Annual Report*, pp. 81-82 and Table 5.2, p. 88.

12. Northwest Energy Policy Project, *Energy Futures Northwest*, pp. 59-60.

13. See, e.g., NRDC Alternative Scenario, pp. 122-23, or Department of Energy, *Oregon's Energy Future*, pp. 59-60, where a production cost of 20-30 mills/kwh is projected for 1985 operation, a figure comparable to the generation

cost of thermal. Cf. Bonneville Power Administration, Branch of Power Resources, "Progress Report on Pacific Northwest Resource Assessments," January 1979, p. 5.

14. Richard Hellman, *Government Competition in the Electric Utility Industry* (New York: Praeger, 1972), pp. 48-51.

15. On the dimensions of the financial problems of the utility industry in general, see the excellent summary by Mason Willrich, "The Electric Utility and the Energy Crisis," *Public Utilities Fortnightly* (Jan. 2, 1975), pp. 22-28 and (Jan. 16, 1975), pp. 25-34, and Russell J. Profozich, "The Changing Structure of the Municipal Utility Sector," Congressional Research Service.

16. In 1971, 1,000 kilowatt-hours (kwh) of electric power cost $11.45 in Portland, Oregon, in an IOU service area, and $9.08 in Vancouver, Washington, served by Clark County PUD. By 1978, the figures were $27.00 for Portland General Electric ($27.24 for the Oregon service area of Pacific Power and Light, the other IOU serving Portland), but only $11.10 in Clark County. The figures for 1971 are from Federal Power Commission, *Typical Electric Bills,* quoted by Ken Billington, conversation with MEM, November 1978; those for 1978 come from Bonneville Power Administration, *Proposed 1979 Wholesale Rate Increase.*

    Billington observes, however, that the small *retail* rate disparity in 1971 obscures a much larger difference in the contributions to the retail price of electric power. Because of PGE's substantial purchases of secondary hydropower in 1971, the average cost paid by the utility for BPA hydro in that year was 2.14 mills/kwh. Clark County PUD, which paid for its BPA hydro on a firm "requirements" contract, paid 3.2 mills/kwh. Thus, public agency-purchased hydropower cost *more* at wholesale than that sold to the IOU in this case. Another way to say this is that PGE's retail rate was more than five times its federal hydro price, while that of Clark County PUD was slightly less than three times its cost for federal hydro.

    This difference is probably attributable to differences in overall power supplies, tax situations, profit considerations, financing possibilities, and--most controversially--managerial efficiency.

    It should be noted as well that PGE's ability to purchase secondary hydro at lower prices is counterbalanced by the fact that secondary hydropower is not available in all years. Thus over the long run, the IOU wholesale cost of acquiring power is probably higher than that of the public systems, even given the situation prevailing in 1971.

17. Dan Seligman, "Washington Public Power Supply System,", p. 11, quoting an estimate by the office of the Commissioner of Public Utilities, State of Oregon.

18. John W. Ellis, in Senate Hearings, p. 161.

19. The notion of "rural" customer has been uncomfortably vague in much of the discussion. The issue is important because of the rapidly growing use of irrigation in the arid interior areas of the Northwest: according to the BPA Role EIS, "about 7.5 million acres of cropland are being irrigated." Appendix A, p. III-143.

    The 1978 Jackson bill limited farm consumption to an irrigation load less than 100 horsepower. This is a power level of 75 kilowatts. Assuming a five-month long irrigation season, with intensive use of irrigation at the maximum permitted level, this rule would limit farmers to an annual average level of 20,000 kilowatt-hours per month, about twenty times the consumption level of a residential customer.

    Under pressure from agricultural interests in the interior, the four Northwest governors suggested raising the "small farm" limit in their amendments to the 1979 Jackson bill. John V. Evans, Victor Atiyeh, Thomas L. Judge, and Dixy Lee Ray, letter to Henry M. Jackson, April 11, 1979, attachment 10. A low-cost block of 400 horsepower for irrgation was authorized in the Senate-passed version of the Jackson bill.

20. Secondary power production in 1975 amounted to over 1,000 average megawatts, while in the drought year of 1973 only 12 average megawatts were generated. BPA Role EIS, Appendix C, Attachment B, Table III. (An "average megawatt" is an amount of energy produced or consumed which, when averaged over a year, is equivalent to a steady output of one megawatt.)

21. BPA Role EIS, Appendix C, p. IV-116.

22. William G. Hulbert, Jr., letter to Representative Lloyd Meeds, September 8, 1977 (an attachment to the so-called Snohomish Amendment to the PNUCC bill). Chart 2-A estimates secondary benefits will increase from $70 million to

$89 million in 1978-80 to $180 million by 1989-90. This is a significantly larger estimate than the conservative figure quoted in the text.

23. BPA Role EIS, Appendix C, p. IV-115.

24. Bonneville Power Administration, *1977 Annual Report,* p. 22.

25. City of Seattle, City Council Resolution 25783, adopted April 3, 1978.

26. The National Energy Conservation Policy Act of 1978, one of the five components of the National Energy Act, provides that utilities must "arrange for...installation or financing" of conservation measures, although "utilities are prohibited from directly installing such materials or actually making loans." See U.S. Department of Energy, *The National Energy Act,* DOE OPA-0003, part IV, p. 1. This legislation establishes a policy, but does not clear the path entirely. The constitutional credit-lending limit will be voted on in the November 1979 general election in Washington State.

27. City of Seattle, OPP, "Regional Power Planning Study: Costs and Power Allocations for Seattle," p. 26.

28. See the discussion in Donna Lee Klemka, "Pacific Northwest Electrical Energy Planning," pp. 119-20.

29. Harry V. Helton, in Senate Hearings, p. 151.

30. Reserves are generally carried by individual utilities for their own generating resources, and by Bonneville for federal facilities. BPA also provides reserves for unanticipated load growth for all the utilities in the BPA service area. This "load growth reserve" is equal to half the growth in utility load from year to year--315 average megawatts in 1977-78. (See PNUCC, Blue Book for 1977, Table 2.) In addition, BPA supplies reserves for the net-billed thermal plants.

31. Earl Gjelde, assistant power manager, Bonneville Power Administration, interview by DLK, November 1977.

32. This estimate of the size of one of the IF-1 quartiles is taken from Eric Redman, comments on a draft study by the Washington Public Interest Research Group, University of Washington, July 14, 1978 (copy on file at the office of Professor Lee, University of Washington), p. 1. Note that

650 average megawatts is roughly two years' worth of load growth reserves.

33. Such standby resources are already included in regional planning under the Pacific Northwest Coordination Agreement of 1964, although their existence is usually *pro forma*. As former BPA administrator Don Hodel admitted, "if it rains, I don't have to run them. So I can put combustion turbines in my forecast...and then pray for rain. And if it doesn't rain I'm going to buy the power someplace; I'm not going to run...turbines...." Interview by KNL, Portland, February 1978.

34. Eric Redman, conversation with KNL, November 1978, observes that there is surprisingly little case law defining the legal responsibilities of utilities to meet new customer loads.

35. In a study of the period 1947-62, Richard F. Fenno concluded that BPA was weak in both public support and congressional support, when compared to other federal agencies. See *The Power of the Purse* (Boston: Little, Brown, 1966); cited in Francis E. Rourke, *Bureaucracy, Politics, and Public Policy*, p. 27.

36. Theodore J. Lowi, "American Business, Public Policy, Case Studies, and Political Theory," *World Politics*, pp. 677-715, distinguishes between "regulatory" and "distributive" policies, pointing out that distinctive patterns of political activity are associated with each type of policy in American political behavior. Regulatory policies characteristically produce interest-group conflict, while distributive policies have relied on "mutual non-interference" to aggregate sufficient legislative support to gain passage. The tenor of debate over public preference has been, as predicted, dominated by the activity of interest groups, including the American Public Power Association, the trade association which lobbies on behalf of public power in Washington, D.C. (See the discussion in Chapter VI.)

37. Ken Billington, executive director, Washington Public Utility Districts Association, interview by DLK, August 1977.

38. John W. Ellis, Senate Hearings, p. 163. *Cf*. The exchange between Ellis and Congressman John Dingell, House Commerce Hearings, pp. 374-76.

39. See Willrich, "The Electric Utility and the Energy Crisis."

40. Robert Short, in House Commerce Hearings, p. 242.

41. With the recent exception of the agreement by Puget Sound Power and Light to aid in the financing of local government services associated with the construction of the Skagit Nuclear Project, utilities have rarely assisted communities with the impacts of power plant development.

42. Senate Report, p. 66.

43. Oregon Department of Energy, *Oregon's Energy Future, Second Annual Report,* pp. 75-94.

44. Richard Timm and Joseph Nadal, Testimony on proposed specific standards for the determination of economic prudence under Rule 345-75-025(1)(b), to the Oregon Energy Facility Siting Council, pp. 60-61, 74-75.

45. U.S. House of Representatives, "A BILL To establish a regional energy corporation...," H.R. 5862, 95th Cong., 1st Sess., March 31, 1977, the "Weaver bill."

On May 21, 1979, Weaver introduced a revised version, numbered H.R. 4159. This draft follows the form and language of the Jackson bill closely, but adds language that would bar further development of nuclear power in the region, as well as a number of less inflammatory changes.

46. For an account of Phase 2's institutional arrangements, see Chapter IV and BPA Role EIS, Appendix A, pp. IV-97 to IV-104.

47. Bernard Goldhammer, "Hydro-Thermal Power Program of the Pacific Northwest, Round 3."

48. Bonneville Power Administration, "Study Paper for Regional Power Planning," p. 5.

49. Ibid., p. 6.

50. Ibid., pp. 7-8 and Attachment A describe the three-tiered rate structure.

51. *NRDC Alternative Scenario.* A summary of the Alternative Scenario was included as Attachment A of the BPA Role EIS, Part 1.

52. Oregon State Legislative Assembly, S.B. 320, Sec. 43.

53. See Eugene Bardach, *The Implementation Game* (Cambridge: MIT Press, 1977), Chapters 8, 9, for a discussion of how delays can cascade.

54. Norman A. Stoll. "Pros and Cons of Proposed Regional Power Supply Program as Developed to Date," p. 15.

55. The effect of the bill on the formation of new public entities, including DRPA, was never spelled out clearly. It has been an unspoken assumption of the attempts to pass federal legislation that the problems of the region's utilities would be resolved sufficiently to obviate the need to form new public power agencies.

56. PNUCC bill, Sec. 4(h), in Senate Hearings, p. 12.

57. Neil Goldschmidt, "Pacific Northwest Utilities Energy Program," p. 25.

58. Francis J. Ivancie, press statement, Portland Department of Public Utilities, November 12, 1976 (copy on file at the office of Donna L. Klemka, Seattle City Council staff), p. 2.

59. *City of Portland, Oregon, et al.* v. *Donald Paul Hodel and James Schlesinger*, Civ. No. 77-928 (D. Ore.) Complaint for Declaratory Relief and Mandate under the Bonneville Project Act, the Due Process Clause and the Administrative Procedures Act; and Civ. No. 77-929 (D. Ore.) Complaint for Injunctive Relief and Declaratory Relief.

    The claims entered by Portland are broader than those of DRPA. DRPA attempts to be consistent with the preference language of the Bonneville Project Act, since a new public utility--serving only domestic and rural customers--would be created. The Portland litigation claims this reorganization is rendered superfluous by the mandate already contained in the language of the act. Operation of DRPA, moreover, would affect the allocation of federal hydro and with it, the costs faced by Oregon's IOUs and all the public systems in the region. Portland's claims, if implemented, would change the costs of the IOUs outside Oregon as well.

60. See, e.g., Ed Mosey, "Governors Split Over Power Plan," Portland *Oregonian* (April 12, 1978); Steven K. Wagner, "NW Governors Clash With Ray Over Power," Seattle *Post-Intelligencer* (April 12, 1978).

61. Hulbert, letter to Representative Lloyd Meeds, September 8, 1977.

62. Activation of these national interest groups put Northwest public power leaders in the awkward position of defending the PNUCC bill's dilution of the preference policy. The initial stage of this defense was undertaken by Ken Billington, then executive director of the Washington Public Utility Districts Association and a highly respected spokesman for public power. In an unpublished paper entitled "Preference and the Proposed New Pacific Northwest Joint Regional Power Program," Billington concluded that "No precedent is established under this proposal which could be used in other regions of the nation to abrogate or destroy the right of preference to federal power by local preference utility customers in such regions." (p. 11)

    At the time, Billington was preoccupied with lobbying before the Washington State legislature for the PUD Association. By the time the state legislature wound down in May 1977, much of the PNUCC draft had been developed. Despite his support of the PNUCC bill, Billington was disturbed that the language of the draft seemed to put an inordinate burden of cost and risk on publicly owned utilities. (Interview by DLK, Seattle, September 1977).

    But the key issue was the one raised by the Snohomish Amendment: the permanent provision of preference power to IOU customers, to reduce the painful rate disparities.

63. American Public Power Association, Legislative and Resolutions Committee, *Pacific Northwest Power Plan.*

64. Ibid., p. 4.

65. Stoll, "Pros and Cons of Power Supply Program," pp. 14, 15.

66. Goldschmidt, "Pacific Northwest Utilities Energy Program," pp. 12, 19.

67. Stoll, "Pros and Cons of Power Supply Program," p. 26.

68. The firm power sold to the direct-service industrial customers in 1977-78 is estimated at 2,495 average megawatts by the industrial customers' attorney. (Eric Redman, letter to Joan Hohl and Kai Lee, with attachments, July 14, 1978). The power delivered to residential customers of privately owned utilities is roughly similar; this may be

estimated as follows. The 137.8 billion kilowatt-hours of electric energy generated in fiscal 1977 constitute an average power rate of 15.7 megawatts. In the region as a whole, just over 32 percent of the electricity is consumed by residential users; we assume this figure holds in each of the three Northwest states of Idaho, Oregon and Washington. There is a minor inaccuracy in this estimate, because Montana is excluded, and all of Idaho is included, though a major part of the state is not in the West Group Area which, for utility purposes, comprises the Pacific Northwest.) The consumption of electric power by state, and the sales by investor-owned utilities in each state are shown below:

	Idaho	Oregon	Washington
percentage of total electricity sales (1975)	12	29	59
percentage served by investor-owned utilities (1976)	86	63	23
average power delivered to residential customers of IOUs (megawatts)	520	920	680

The total in the last line is approximately 2,100 average megawatts. (Data from BPA, *1977 Annual Report*, p. 20 and Role EIS Summary Report, p. 7; Northwest Energy Policy Project, *Energy Futures Northwest*, final report, 1978, pp. 42-45, from which the first line of the table is inferred; and U.S. Department of Energy, Office of the Regional Representative, Region X, "Regional Electric Power Planning Issues in the Pacific Northwest," June 1978, p. 6, from which the second line in the table is taken.)

69. Dean Katz, "Regional Power Bill Finally Surfaces in Senate," Seattle *Times* (July 29, 1979), p. A8.

70. This discussion of the Jackson bill follows the Senate Report prepared by the Committee on Energy and Natural Resources. After the bill was reported out of committee, three amendments were added on the floor of the Senate. Two were offered by Energy Committee members and amounted to minor redrafting. A third amendment, by Sen. Warren G. Magnuson, urged further development of hydropower, especially the ample resources of British Columbia.

71. Senate Report, Sec. 4(e), p. 3.

72. Ibid., Sec. 5 preamble; pp. 4, 5.

73. Irrigation pumping was, at the behest of Idaho's Sen. James McClure, given an extraordinary subsidy, with the first 400 horsepower of pumping included in the "residential" rate. See Sec. 3(o), Senate Report, p. 3.

74. The rate test is delineated in Sec. 7(b) and the IOU pull-back language in Sec. 5(b). An analysis of the projected impact of the rate test, prepared by Bonneville, is included as Appendix B of the Senate Report; see pp. 5, 9-10, 56-79.

75. See ibid., Sec. 3(c), 4(e)-(f), 6(a)-(b), 6(h), pp. 2-4, 6, 8.

76. On the role of the administrator, see ibid., Secs. 4(a)-(b), 6(c), 9(b), pp. 3, 6-7, 11-12, 33. The council is funded by a surcharge on BPA power sales; a ceiling of roughly $2 million per year is set in Sec. 4(c); ibid., p. 3. Quote from Larry Hittle, BPA division of power management, interview by KNL, Portland, July 1979.

77. Senate Report, Sec. 3(c), p. 3.

78. Ibid., Sec. 4(h), p. 4.

79. Ibid., p. 24.

80. As it "marked up" or amended the bill, the Senate Energy Committee debated who should determine the consistency of the administrator's actions with the plan. Sec. 4(d) (ibid., p. 3) authorizes the regional council to make a finding of inconsistency by majority vote--that is, with three of the four state representatives voting against a reluctant BPA chief. It seems doubtful that even a powerful Bonneville administrator would deviate so far from regional consent as to require this sort of reprimand. Of somewhat greater import, the language regarding the definition of consistency implicitly measures the independence accorded the administrator. By requiring an extraordinary three-state majority, the Senate endorsed a strong BPA role. "You've got a regional entity which does not represent any *bona fide* political subdivision," a Senate committee staff member said. "There is no regional government." Joel Merkel, interview by KNL, Washington, D.C. June 1979.

81. The bill requires the Bonneville administrator to state his "findings" regarding the suitability of major resource acquisitions. Sec. 6(c), Senate Report, p. 6. This terminology permits judicial review under the test that the administrator's decision must not be arbitrary and capricious, a standard substantially looser than that requiring a decision after a hearing on the record under the Administrative Procedures Act. The hearing on the record essentially requires evidence to back up the administrator's decision, not simply an indication that he has considered alternative courses of action. The Administrative Procedures Act language was originally called for in the amendments submitted by the four Northwest governors. See John V. Evans, Victor Atiyah, Thomas L. Judge, and Dixy Lee Ray, letter to Hon. Henry M. Jackson, April 11, 1979, attachment, p. 7, and Roger Beers and Terry R. Lash, "Changes Recommended by the Natural Resources Defense Council in S. 885 and H.R. 3508," unpublished memorandum, June 20, 1979 (copy on file at the office of Professor Lee).

82. Beers and Lash, letter to Senator Frank Church, June 15, 1979, p. 1.

83. Senate Report, Sec. 9(e), p. 13.

84. Andy Athy, counsel to the House Subcommittee on Energy and Power, interview by KNL, Washington, D.C., June 1979.

85. Representative Al Swift, Democrat of Washington, interview by KNL, Washington, D.C., June 1979.

86. Ibid.

87. Representative Norm Dicks, Democrat of Washington, interview by KNL, Washington, D.C., June 1979. The total, not annual, cost of the WPPSS projects is now approaching $12 billion.

88. Tim Cottrell, legislative assistant to Representative Mike Lowry, Democrat of Washington, interview by KNL, Washington, D.C., June 1979.

89. Athy to KNL.

Chapter VI. *Institutional Analysis*

1. Perry Miller, "The Responsibility of Mind in a Civilization of Machines," *American Scholar,* reprinted in *Changing Attitudes Towards American Technology,* edited by Thomas Parke Hughes, p. 72.

2. Theodore J. Lowi, "American Business, Public Policy, Case-Studies, and Political Theory," *World Politics,* pp. 677-714; Robert H. Salisbury, "The Analysis of Public Policy: A Search for Theories and Roles," in *Political Science and Public Policy,* pp. 151-178.

   Lowi developed his classification of public policy types further in "Decision Making vs. Policy Making: Toward an Antidote for Technocracy," *Public Administration Review* 30, pp. 314-325 and "Four Systems of Policy, Politics, and Choice," ibid., 32, pp. 298-310. The Lowi classification is criticized in James Q. Wilson, *Public Organizations,* Chap. 16. A good discussion with a somewhat revised typology, is contained in Michael T. Hayes, "The Semi-Sovereign Pressure Groups: A Critique of Current Theory and an Alternative Typology," *Journal of Politics,* p. 688.

3. Lowi, "American Business...," p. 688.

4. Ibid.

5. The following table summarizes the classification in ibid., and Salisbury, "Analysis."

Ideal Type	Characteristics
Distributive	Government distributes public property or funds, under conditions where "decisions can be made without regard to limited resources." (Lowi, p. 690).
	Examples: defense procurement; tax exemptions for special interests; water projects, including Columbia River dams.
	Easily divisible benefits, which can be dispensed unit by unit.
	Loser and recipient need never come into direct confrontation.
	Political relationship: stable

"logrolling" coalition based on shared constituency interests among legislators, not ideological or policy agreement; "mutual non-interference."

Regulatory

Government makes "a direct choice" of whose behavior is to be permitted; though "implementation is firm-by-firm and case-by-case...individual decisions must be made by application of a general rule." (Lowi, pp. 690-691.)

Examples: airline route awards; utility rate structures; pollution control permits and power plant licenses; BPA purchase agreements.

Benefits linked to one another by general rule, so that legal standards become necessary.

Those affected by regulatory constraints or awards form an easily identifiable group with common interests, while those who benefit from constraints or pay for regulatory licensing are normally the general public.

Political relationship: *interest-group bargaining* characterized by unstable coalitions, conducted by leaders whose group interests and personal styles overlap with one another; pluralism.

Redistributive

Government takes from one broad class of individuals and gives to another.

Examples: income tax; Medicare; residential preference.

"The nature of a redistributive issue is not determined by the outcome of a battle over how redistributive a policy is going to be.

	Expectations about what it *can* be, what it threatens to be, are determinative." (Lowi, p. 691.)
	Individual decisions are linked to one another, as in regulatory policy.
	Political process: *class conflict;* power elite.
Self-Regulatory (Salisbury, p. 158)	Government delegates future decision-making authority to a private group.
	Examples: professional licensing; Constitutional amending process; PNEPCO "gating" authority
	Only benefits are conferred, and the costs are borne by the general public.
	Once the policy is enacted, political pressure ceases.
	Political process: non-ideological, *temporary coalition.*

6. Wilson, *Political Organizations,* p. 328, comments that the Lowi-Salisbury scheme remains "ambiguous and incomplete. It is hard to distinguish the three arenas of power from one another except in extreme cases; there are broad areas of power that seem to fit nowhere in the scheme; and there are important changes over time in the way groups behave."

   As Figure 13 points out, more than one type of policy has been declared in the Northwest power arena. Consistent with the Lowi-Salisbury approach, patterns of political conflict have been varied, but unfortunately the simple associations between policy type and political configuration are muddled by the coexistence of different policy types. In addition, the groups active in the regional power arena have changed dramatically over time, as we indicated in Chapter IV.

7. The vertical dimension is adapted from ibid., pp. 327-337, and the horizontal dimension is taken from Salisbury, "Analysis," pp. 163-174. The choice of dimensions for this anal-

ysis is deliberately eclectic: the combination of economic considerations, balancing short run and long term emphases in choice, together with political considerations, characterized by a tension between self-governance and regional interdependence, seem to us to capture most clearly the issues at stake in the Northwest case. Given the theoretical and methodological limitations of the Lowi approach, theoretical consistency seems less important than clarity of application.

8. The iron triangle or "subgovernment" has inspired an enormous literature in political science. See, *e.g.*, Grant McConnell, *Private Power and American Democracy*. The account we develop here follows the sociological analysis of Hugh Heclo, *A Government of Strangers*.

9. Heclo, *Strangers*, p. 47.

10. Ibid., p. 111.

11. Anonymous interview by KNL, Washington, June 1979.

12. Don Hodel, interview by KNL, Portland, February 1978.

13. See footnote 5.

14. See, for instance, the thoughtful analysis of the tension between material and social motives in Sheldon Wolin, *Politics and Vision*, pp. 406-414. Compare with the discussion of "conflicting rationalities" in Ernst B. Haas, *The Obsolescence of Regional Integration Theory*, pp. 87-88, and the broad review contained in Langdon Winner, *Autonomous Technology*.

15. See Galbraith's trilogy on the modern economy, *The Affluent Society; The New Industrial State;* and *Economics and the Public Purpose*. Bell's *The Coming of Postindustrial Society* provides a broadly suggestive account of the directions of evolution in the social order that he projects.

16. In fact, the "track record" of the *Energy 1990* forecast has been rather good, suggesting that the projections on which the Seattle City Council acted have been borne out. In the 1976-79 period actual electric energy demand has been consistently below the Energy 1990 forecast, and the amount of shortfall has been growing. (See Randy Revelle, letter to Congressman John Dingell, in House Commerce Hearings, p. 367, for a comparison of demand projections.)

17. Edward C. Banfield, *The Unheavenly City Revisited*, p. 261. Banfield summarizes the shortcomings of reform thus:
    °Usually, a reform is suggested without any notion of how practically to implement it (conservation is frequently attacked on this ground);
    °Often reforms require extensive motivation of individual behavior, but such exercise of government authority is politically infeasible (mandatory conservation);
    °Many reforms do not properly analyze the social conditions responsible for the pathologies to be cured (focusing on frivolous residential uses of electricity, when expansion of irrigation may have a much larger, more readily corrected effect on load growth);
    °Reforms are usually wildly costly in relation to the benefits obtained (environmental restrictions are attacked for this reason).

    Moreover, feasible changes are politically unacceptable:
    °Interest groups can veto reform (defense of public preference);
    °Widely shared benefits lack articulate advocates (who will carry the case for the fish in the Columbia far enough?);
    °Public opinion tends toward "mass extravaganzas" that are poorly thought out, morally insensitive, and involve extensive danger of doing substantial harm in the pursuit of uncertain or unobtainable good (redistributing public power to rural and domestic customers, which will only temporarily stem the tide of rising cost).

18. Ibid., p. 277.

19. Robert D. Thomas, "Intergovernmental Coordination in the Implementation of National Air and Water Pollution Policies," *Public Policy Making in a Federal System*, Chap. 6.

20. On the difficulties of representation, see Paul E. Peterson, "Focus of Representation: Participation of the Poor in the Community Action Program," *American Political Science Review*, pp. 491-507, and compare Frances Fox Piven and Richard C. Cloward, *Regulating the Poor*, Chap. 9, 10.

    On the evolution of federalism, see Richard H. Leach, *American Federalism*.

21. As Robert C. Wood has observed, there is already an endemic tendency for policy making to drift downward in the federal bureaucracy, until policy choices are made by persons far removed from political accountability; see "When Government Works," *Public Interest*, pp. 39-51.

22. Don Hodel, "The Prophets of Shortage (Part 2)."

23. Charles L. Schultze, *The Public Use of Private Interest,* p. 7.

24. For an insightful summary of the bureaucratization of political judgment, see Heclo, *A Government of Strangers,* Chap. 2; Heclo also describes the process of politicizing the bureaucracy mentioned earlier.

25. The regulatory function of government purchases has been explored recently by writers critical of the American economic order. Perhaps the most original thesis is that of James O'Connor, *The Fiscal Crisis of the State,* who analyzes government participation in the private economy as an attempt to shore up the inherent weaknesses of a capitalist system.

    Evidence from a drastically different perspective may be found in J. Ronald Fox, *Arming America,* p. 450. In his study of defense procurement, Fox suggests that firms manufacturing products for a single large buyer face extraordinary problems in maintaining their levels of capital investment. This analysis is directly opposite that of O'Connor, for whom the activities of government are designed precisely to advance capital accumulation.

    Note that the situation in the Pacific Northwest is ambiguous on this point. Government investment in the past seems to have displaced investment by the utilities, but Bonneville purchase would facilitate investment by utilities in the future. In both cases, however, it is difficult to estimate the levels of investment activity that would have prevailed in the absence of government participation. The issue is not academic, given the controversy which surrounds the need for additional generating resources and the possibility of enforcing conservation practices through the threat of shortages.

26. Haas, *Obsolescence,* p. 18.

27. The distinction drawn here parallels one used by Wolin to characterize contending schools of thought in organizational sociology; *Politics and Vision,* pp. 409-410.

28. William A. Gamson, "Stable Unrepresentation in American Society," *American Behavioral Scientist* (November-December 1969), p. 20.

29. Charles A. Lave and James G. March, *An Introduction to Models in the Social Sciences* in Chap. 3 discuss alternative criteria for evaluating social process models; "truth," or success in explaining phenomena, is not the only criterion. Usability in political action--which resembles in some ways the criterion of "justice" discussed by Lave and March--is not often found in social analysis.

30. Ernst B. Haas, *Beyond the Nation-State,* p. 88.

31. Ibid., pp. 94-95. This argument draws upon the discussion in Philip Selznick, *Leadership in Administration*, which we explore in more detail in Chapter VII.

32. The phrase is taken from Heclo, *Strangers,* p. 6. See also Charles Perrow, *Complex Organizations: A Critical Essay* Chap. 1, for a defense of bureaucratic organization.

33. Understandably, there was a burst of optimism that regional integration would supersede the nationalist competition that had ravaged the world in two global conflicts within a generation. A sober, sophisticated appraisal of the possibilities of regional integration is Haas' *Beyond the Nation-State*. We draw upon Haas' ideas in Chapter VII, in our analysis of the Northwest integration problem. We are indebted to Professor Hayward Alker of MIT for reminding us of the relevance of regional integration theory to the problem at hand.

34. See Roger G. Noll, *Reforming Regulation,* Chap. 4, for a brief summary of the literature on utility regulation.

35. For an analysis of the symbiosis between multinational corporations and U.S. foreign policy, see Robert Gilpin, *U.S. Power and the Multinational Corporation,* esp. Chap. 6.

36. Ibid., p. 259.

37. Haas, *Obsolescence,* Chap. 5.

38. Robert O. Keohane and Joseph S. Nye, *Power and Interdependence,* Chap. 2.

Chapter VII. *Institutional Alternatives*

1. Elting E. Morison, *From Know-How to Nowhere* (New York: Basic Books, 1974), p. x.

2. *Power Review 77*, pp. 68-69.

3. Bonneville Power Administration, *1978 Annual Report*, (Portland: BPA, 1979), p. 38.

4. Senate Report, pp. 65 (value of regional reserves" provides and estimate of capital costs), 71 (line 70), 73 (line 70).

5. See *NRDC Alternative Scenario*.

6. Don K. Price, *The Scientific Estate* (New York: Oxford University Press, 1978), p. 162.

7. This discussion is adapted from J. David Greenstone and Paul E. Peterson, *Race and Authority in Urban Politics* (New York: Russell Sage Foundation, 1973), pp. 100-109; and Charles Tilly, *From Mobilization to Revolution* (Reading, Mass.: Addison-Wesley, 1979), Chaps. 3, 4. We are indebted to Professor David Olson and Professor Margaret Levi for suggesting these sources.

8. Compare Stuart A. Scheingold, "The Community in Perspective: Public Policy and Political Structure," *Annals American Academy of Political and Social Sciences* 440 (1978), pp. 156-167.

9. See Matthew A. Crenson, *The Un-politics of Air Pollution* (Baltimore: Johns Hopkins University Press, 1971).

10. Figure 21 is influenced by Greenstone and Peterson, *Authority* pp. 264-293, and Tilly, *Mobilization*, Chap. 3. A related typology is developed in great detail by John Gerard Ruggie, "International Responses to Technology: Concepts and Trends," *International Organization* 29 (1975), pp. 557-583. Ruggie's classification suggests a large number of alternative institutional forms, and our work has been enriched by his inventory of cooperative relations among sovereign governments.

11. Ernst B. Haas, *Beyond the Nation-State* (Stanford: Stanford University Press, 1966), p. 6. The term "purely rational" here refers to decisions made under circumstances in which both ends and means are agreed upon, as in the upper-left entry of Fig. 20. (We are indebted to Professor Hayward R.

Alker, Jr., for pointing out the relevance of Haas' analysis to the Pacific Northwest case.)

12. Philip Selznick, *TVA and the Grass Roots* (Berkeley: University of California Press, 1949); Deborah Shapley, "TVA Today: Former Reformers in an Era of Expensive Electricity," *Science* 194 (1976), pp. 814-818.

13. On the disappointing capacity of the Delaware River Basin Commission to achieve--or even to conceive--meaningful water pollution control, see Bruce A. Ackerman and Susan Rose-Ackerman, James W. Sawyer, Jr., Dale W. Henderson, *The Uncertain Search for Environmental Quality* (New York: Free Press, 1974), Chap. 13.

14. Selznick, *TVA*, esp. pp. 69ff, 250-266.

15. Selznick, *Leadership in Administration* (New York: Harper and Row, 1957), pp. 14-17. For a related and exceptionally clear discussion, see Francis E. Rourke, *Bureaucracy, Politics, and Public Policy* (Boston: Little, Brown, 1969), Chap. 4.

    Selznick's ideas were applied to the problems of international cooperation by Haas in *Beyond the Nation-State;* our formulation draws heavily on Haas' analysis.

16. Selznick, *Leadership,* pp. 22-28.

17. Ibid., pp. 74-82.

18. Greenstone and Peterson, *Authority,* p. 269.

19. Selznick, *Leadership,* p. 126.

20. BPA, *1978 Annual Report,* pp. 54-55.

21. Charles Perrow, *Complex Organizations: A Critical Essay* (Glenview, Ill.: Scott Foresman, 1972), p. 191.

22. The term was coined by John Kenneth Galbraith, *Economics and the Public Purpose* (New York: New American Library, 1973), Chap. 24.

23. Meg Greenfield, "Thinking Small," *Newsweek* (April 24, 1978), p 112.

24. In the classification developed by Ruggie in "Responses," our proposals would be referred to as a "regime."

25. Such foreshortening is, of course, far from unique. Haas concluded that the growth of international organizations was not accompanied by any weakening of nationalism; see *Nation-State*, p. 459.

26. The international experience indicates that factors external to international organizations and national governments tend to be the principal forces pressing for international cooperation. See Haas, ibid.

27. Ibid., pp. 94-95.

28. Charles McKinley, *Uncle Sam in the Pacific Northwest* (Berkeley: University of California Press, 1952), pp. 189ff, documents the decline of public power leadership.

29. J. Serge Taylor, letter to KNL, September 1978.

30. Early in 1979, Robert Murray became superintendent of Seattle City Light, the region's largest public utility, after a bruising confirmation battle. His appointment was fiercely resisted by the Technocratic Elite because of his previous sympathy with utility critics. Murray's fate as superintendent will be an important--if not necessarily accurate--indicator of the institutionalization of precarious values in the region.

31. Alan A. Altschuler and Robert W. Curry, "The Changing Environment of Urban Development Policy--Shared Power or Shared Impotence?" *Urban Law Annual* 10 (1975), pp. 3-41.

32. Harvey R. Doerksen, *Columbia River Interstate Compact, Politics of Negotiation* (Pullman, Wash.: Water Research Center, 1972), pp. 177ff, concludes that the divisions of the public power movement were a principal cause of the failure of the Columbia River Interstate Compact--an interstate agreement for water resource development.

33. A similar idea is argued in Anthony Downs, "Up and Down with Ecology--the Issue-Attention Cycle," *Public Interest* 28 (Summer 1972), pp. 38-50.

34. Thus, institutionalization of the public interest can substitute for legislative declarations of public purposes. For a criticism of legislation as policy making, see Theodore J. Lowi, *The End of Liberalism* (New York: Norton, 1969), Chap. 5, 10. But Galbraith, *Economics*, Chap. 29, recommends reliance on Congress for the definition of public purposes.

35. The chairperson should serve a full-time, three-year term. If he has less than three years left in his five-year term at the time of appointment, his term as a member of the council should be extended.

36. We are grateful to Randall W. Hardy for pointing this out. Letter to KNL, August 1978.

37. Rourke, *Policy*, p. 24.

38. Compare Murray Edelman, *The Symbolic Uses of Politics* (Urbana: University of Illinois Press, 1964), Chap. 1, 4, 5.

39. 16 U.S. Code 832c(a).

40. Don Hodel, interview by KNL, Portland, February 1978.

41. Selznick, *TVA*, pp. 265-266.

42. Charles L. Schultze, *The Public Use of Private Interest* (Washington: Brookings Institution, 1977), p. 12.

43. Ackerman et al., *Uncertain Search*, p. 220.

44. A Myrick Freeman III, Robert H. Haveman, and Allen V. Kneese, *The Economics of Environmental Policy* (New York: Wiley, 1973), Chap. 5.

45. An influential discussion of public goods is Mancur Olson, *The Logic of Collective Action* (Cambridge: Harvard University Press, 1965).

46. See Schultze, *Public Use*, passim.

47. Measures to manage peak demands include
    --interrupting aluminum-plant loads during peak hours;
    --adding a peak-demand charge to irrigation pumping;
    --developing additional upstream storage whose principal purpose is authorized to be fish protection, instead of hydropower;
    --developing small, flexible generating plants, including pumped storage, to lower the peaking burden on the mainstem hydro plants.

    Because of the extremely low cost of hydro generation there will always be economic pressure to raise hydro peaking--and with it, destructive fluctuations in river level. Thus the need to protect fishery values by non-market methods. The high value of hydro means, moreover, that funding of protective measures will not impair "bargain basement" hydro costs. A surcharge of only half a mill yields $35 million annually.

48. David J. Bardin, in House Commerce Hearings, p. 43. Both this and PIP are extensions of the Seattle rate credit proposal, which would provide economic benefits (through adjustments in wholesale power rates via rate credits) to utilities which saved beyond the cost-effective standard. In economic theory low-cost hydro should be allocated on the basis of inverse elasticity--those least able to adjust their consumption or efficiency should get the largest fraction of the hydro. That way, price incentives would bring about savings from those able to change. Such a principle has a decidedly poor chance of adoption in circumstances where residential users may show considerable elasticity of demand as compared to industry. We are grateful to Professor M. Granger Morgan for a helpful discussion on this point.

49. On the general problems of implementing public policy, see Eugene Bardach, *The Implementation Game* (Cambridge: MIT Press, 1977).

50. See Ackerman et al., *Uncertain Search*, esp. Chap. 13.

51. See, "Notice of Proposed Policy and Formula to Guide Allocation of Firm Electric Energy...," *Federal Register*, October 5, 1979.

52. BPA Role EIS, Part 1, p. VI-7.

53. The assessment has been a closely held document. A draft summary, "Key Elements of the Assessment of Arrangements," is included as Appendix F of Dan Seligman, "The Washington Public Power Supply System" (unpublished paper, University of Washington, Institute for Environmental Studies, December 1978); copy on file at the University of Washington.

54. Bonneville Power Administration, "Allocation of Bonneville Power Resources," discussion paper, June 2, 1978.

55. See, for example, Bonneville Power Administration, *Pacific Northwest Long Range East-West Corridor Study*, Phase I, Part A (Rocky Mountains), December 1977 draft working papers (Portland: BPA, 1977).

56. For an account, see Doerksen, *Interstate Compact*.

57. For a discouraging appraisal of BPA's ability to oversee WPPSS, see Comptroller General of the United States, *Impacts and Implications of the Pacific Northwest Power Bill*, EMD-79-105 (Washington: U.S. General Accounting Office, September 4, 1979), pp. 7-13.

58. On the importance of providing analytic capabilities for public debates in technically complex areas, see Robert H. Socolow, "Failures of Discourse," in *Boundaries of Analysis*, ed. Harold A. Feiverson, Frank W. Sinden, and Socolow (Cambridge, Mass.: Ballinger, 1976).

59. For an excellent and insightful appraisal of NEPA as a means of institutionalizing scientific analysis in federal policy, see J. Serge Taylor, "Environmentalists Inside the Bureaucracy," Ph.D. dissertation, University of California, Berkeley, 1978.

60. John Lukacs, *The Passing of the Modern Age* (New York: Harper Torchbooks, 1970), p. 138.

Bibliography

Ackerman, Bruce A.; Rose-Ackerman, Susan; Sawyer, James W., Jr.; Henderson, Dale W. *The Uncertain Search for Environmental Quality*. New York: Free Press, 1974.

Altschuler, Alan A., and Curry, Robert W. "The Changing Environment of Urban Development Policy--Shared Power or Shared Impotence?" *Urban Law Annual* 10 (1975):3-41.

American Public Power Association. "Tacoma's Low Power Costs Cited in Survey by TV Station." *Public Power Weekly*, January 16, 1978.

_______, Legislative and Resolutions Committee. *Pacific Northwest Power Plan*. Resolution adopted January 25, 1978. Washington, D.C.

Anschell, Cleveland. "Government City, Richland, Washington." *Urban Trends*, Rainier National Bank Newsletter 1 (1978):6-8.

Athy, Andy. Counsel to House Subcommittee on Energy and Power. Interview by KNL. Washington, D.C., June 1979.

Bamesberger, Lee. Department of Chemical Engineering, Washington State University, Pullman, Wa. Telephone interview by KNL. July 1976.

Banfield, Edward C. *The Unheavenly City Revisited*. Boston: Little, Brown and Co., 1974.

Bardach, Eugene. *The Implementation Game*. Cambridge: MIT Press, 1977.

Bardsley and Haslacher, Inc. "Electric Energy and the Environment." Study report for PNUCC, November-December 1975. Copy on file at offices of Culp Dwyer Guterson & Grader, attorneys. Seattle, Wa.

Beers, Roger, and Lash, Terry R. *Choosing an Electrical Energy Future for the Pacific Northwest: An Alternative Scenario*. Palo Alto, Calif.: Natural Resources Defense Council, Inc., January 1977.

_______. Letter to Senator Frank Church, June 15, 1979.

_______. "Changes Recommended by the Natural Resources Defense Council in S. 885 and H.R. 3508." Memorandum, June 20, 1979. Copy on file office of KNL, Institute for Environmental Studies, University of Washington, Seattle, Wa.

Bell, Daniel. *The Coming of Post-Industrial Society*. New York: Basic Books, 1973.

Berlin, Edward; Cicchetti, Charles J., and Gillen, William J. *Perspective on Power*. Cambridge, Mass.: Ballinger, 1974.

Bessey, Roy F. *Pacific Northwest Regional Planning--A Review*. Bulletin No. 6, Division of Power Resources, Department of Conservation. Olympia, Wa.: State Printing Plant, 1963.

Bickel, Alexander. *The Least Dangerous Branch*. Indianapolis: Bobbs-Merrill, 1962.

Billington, Ken. Executive Director, Washington Public Utility Districts Assoc. Interview by DLK. Seattle, Wa., August 1977.

_______. "Preference and the Proposed New Pacific Northwest Joint Regional Power Program." Unpublished paper, 1977. Copy on file in office of KNL, Institute for Environmental Studies, University of Washington, Seattle, Wa.

_______. Conversation with MEM, November 1978.

Birmingham, Alison D. "Regional Electric Energy Planning: A Case Study in the Politics of Scarce Resources." Mimeographed. University of Washington, Seattle, Wa., 1976.

Boffey, Philip M., and Wade, Nicholas. "The Nuclear Debate: Clashes in Congress and California." *Science* 191 (1976):47-50.

Bonneville Power Administration. Department of Energy. *A Ten Year Hydro-Thermal Power Program for the Pacific Northwest*. Portland, Ore.: BPA, 1969.

_______. "Net Billing Analysis." Unpublished memorandum. Portland, Ore.: BPA, January 3, 1973.

_______. *About BPA*. Portland, Ore.: BPA, 1974.

_______. "Study Paper for Regional Planning." Portland, Ore.: BPA, November 5, 1976.

_______. *1976 Annual Report*. Portland, Ore.: BPA, 1977.

_______. *The Role of the Bonneville Power Administration in the Pacific Northwest Power Supply System, Including Its Participation in the Hydro-Thermal Power Program*, draft environmental statement, U.S. Department of the Interior DES 77-21, July 22, 1977, in seven volumes. Portland, Ore.: BPA.

_______. *Pacific Northwest Long Range East-West Corridor Study*, Phase I, Part A (Rocky Mountains), December 1977 draft working papers. Portland, Ore.: BPA, 1977.

_______. *1977 Annual Report*. Portland, Ore.: BPA, 1978.

_______. "Allocation of Bonneville Power Resources." Discussion paper. Portland, Ore.: BPA, June 2, 1978.

_______. *Proposed 1979 Wholesale Rate Increase,* draft environmental statement, Department of Energy DOE/EIS-0031-D. Portland, Ore.: BPA, 1978.

_______. *1978 Annual Report.* Portland, Ore.: BPA, 1979.

_______. Branch of Rates. "Analysis of Rate Directives, Including Preference Customer Rate Limit." Draft paper. Portland, Ore.: BPA, February 1979.

_______. Branch of Rates. "Bonneville's Consideration of Ratemaking Standards Under the Public Utility Regulatory Policy Act (PURPA)." Discussion paper. Portland, Ore.: BPA, June 1979.

_______. "Proposed Policy and Formula to Guide Allocation of Firm Electrical Energy and System Reserve Energy from the Federal Columbia River Power System." Discussion paper. Portland, Ore.: BPA, September 27, 1979.

_______. "Notice of Proposed Policy and Formula to Guide Allocation of Firm Electric Energy and System Reserve Energy from the Federal Columbia River Power System and Opportunities for Public Review and Written Comment." Portland, Ore.: BPA, October 5, 1979. Also published in *Federal Register,* October 5, 1979.

Bonneville Project Act. "AN ACT to authorize the completion, maintenance, and operation of Bonneville project for navigation, and for other purposes." Act of August 20, 1937, Ch. 720, 50 Stat. 731, as amended.

Branscombe, James. "Destroy to Save?" *Environment,* September (1975):6-11.

Brewer, William A. Conversations with MEM and KNL. Seattle, Wa.: 1974 and 1975.

Breyer, Stephen, and MacAvoy, Paul W. "The Federal Power Commission and the Coordination Problem in the Electrical Power Industry." *Southern California Law Review* 46 (1973): 661-712.

Burwell, C.C.; Ohanian, M.J., and Weinberg, A.M. "A Siting Policy for an Acceptable Nuclear Future." *Science* 204 (1979):1043-1051.

Cameron, Juan. "Where *Does* the President Stand on Nuclear Energy?" *Fortune* (March 1978):99ff.

Carter, Luther J. "Con Edison: Endless Storm King Dispute Adds to Its Trouble." *Science* 184 (1974):1353-1358.

_______. "Nuclear Initiatives: Two Sides Disagree on Meaning of Defeat." *Science* 194 (1976):811-812.

_______. "SO_2 Emissions Pose Growth Issue." *Science* 202 (1978):30.

City of Seattle. Office of Policy Planning. "Regional Power Planning Study: Costs and Power Allocations for Seattle." Seattle, Wa.: City of Seattle, February 14, 1978.

_______. City Council Resolution 25783. April 3, 1978.

Claussen, Vera Edinger. *Public Power--Rags to Riches.* Seattle: Washington Public Utility Districts Association, n.d.

Comey, David Dinsmore. "Will Idle Capacity Kill Nuclear Power." *Bulletin of the Atomic Scientists* (November 1974):23-28.

Cottrell, Tim. Legislative assistant to Representative Mike Lowry, Democrat of Washington State. Interview by KNL. Washington, D.C. June 1979.

Crenson, Matthew A. *The Un-politics of Air Pollution.* Baltimore: Johns Hopkins University Press, 1971.

Dicks, Norm. Democrat of Washington State. U.S. House of Representatives. Interview by KNL. Washington, D.C. June 1979.

Doerksen, Harvey R. *Columbia River Interstate Compact, Politics of Negotiation.* Washington Water Research Center, Report No. 11. Pullman, Wa.: Water Research Center (1972):127-137.

Downs, Anthony. "Up and Down with Ecology--the Issue-Attention Cycle." *Public Interest* 28 (Summer 1972):38-50.

Dyer, Polly. Interview by KNL. Seattle, Wa., April 1978.

Edelman, Murray. *The Symbolic Uses of Politics.* Urbana: University of Illinois Press, 1964.

Ernst and Ernst. (Consulting firm). *The Energy-Economy Relationship.* Report to Bonneville Power Administration. Portland, Ore.: BPA, 1976.

Evans, John V.; Atiyeh, Victor; Judge, Thomas L.; and Ray, Dixy Lee. Letter to Henry M. Jackson, with attachments. April 11, 1979.

Fadeley, Edward N. "Preference Rights of Publics and Cooperatives." Mimeographed memorandum. Eugene, Ore.: n.d., c. 1974. Copy on file at office of Professor Marion E. Marts, University of Washington.

"Fallout from the California Nuclear Initiative." *Environmental Law Reporter* 6 (1976):10174.

Farris, Martin T., and Sampson, Roy J. *Public Utilities: Regulation, Management, and Ownership.* Boston: Houghton Mifflin Co., 1973.

Federal Power Commission. *Typical Electric Bills.* Washington, D.C.: FPC, 1971.

Foote, Jeffrey P.; Larsen, Alan S.; and Maddox, Rodney S. "Bonneville Power Administration: Northwest Power Broker." *Environmental Law* 6 (1976):831-857.

Fox, Ronald. *Arming America.* Boston: Harvard Business School, 1974.

Freeman, A. Myrick, III.; Haveman, Robert H.; Kneese, Allen V. *The Economics of Environmental Policy.* New York: Wiley, 1973.

Freeman, S. David. *Energy--The New Era.* New York: Vintage, 1974.

Friedlander, Gordon D. "The Story of Bonneville Power: 1937-1968-1987." *Institute of Electrical and Electronic Engineers Spectrum* (November 1968):78-94; (December 1968):61-72.

Galbraith, John Kenneth. *The Affluent Society*. Boston: Houghton Mifflin Sentry, 1969.

_______. *The New Industrial State*. New York: New American Library, 1971.

_______. *Economics and the Public Purpose*. New York: New American Library, 1973.

Gamson, William A. "Stable Unrepresentation in American Society." *American Behavioral Scientist* 2 (November-December 1969):20.

Garman, Gerald. Power Manager of Seattle City Light. Interview by KNL, July 1979.

Gilmore, John S. "Boom Towns May Hinder Energy Resource Development." *Science* 191 (1976):535-540.

Gilpin, Robert. *U.S. Power and the Multinational Corporation*. New York: Basic Books, 1975.

Gjelde, Earl. Assistant BPA power manager. Interview by DLK. Portland, Ore., November 7, 1977.

Goldhammer, Bernard. "Hydro-Thermal Power Program of the Pacific Northwest, Round 3." Memorandum to the direct-service industrial customers of BPA, November 24, 1975. Copy on file in office of KNL, Institute for Environmental Studies, University of Washington, Seattle.

_______. Interview by MEM, Robert G. Walton, and Alison D. Birmingham. Portland, Ore., April 1976.

Goldschmidt, Neil. "Pacific Northwest Utilities Energy Program." Memorandum to Portland City Council. 1977. Copy on file with DLK, Seattle, Wa.

Gordon, Richard L. "The Hobbling of Coal." *Science* 200 (1978): 153-158.

Graham, Frank, Jr. "The Outrageous Mr. Cherry and the Underachieving Nukes." *Audubon* (September 1977):50-67.

Greenfield, Meg. "Thinking Small." *Newsweek*. (April 24, 1978):112.

Greenstone, J. David, and Peterson, Paul E. *Race and Authority in Urban Politics*. New York: Russell Sage Foundation, 1973.

Haas, Ernst. *Beyond the Nation-State*. Stanford, Calif.: Stanford University Press, 1964.

_______. "Is There a Hole in the Whole?" *International Organization* 29 (1975):876.

_______. *The Obsolescence of Regional Integration Theory*. Research Series No. 25, Institute of International Studies. Berkeley: University of California, 1975.

Hardy, Randall W. Letter to KNL. August 1978.

Harte, John, and El-Casseir, Mohamed. "Energy and Water." *Science* 199 (1978):632.

Hayes, Michael T. "The Semi-Sovereign Pressure Groups: A Critique of Current Theory and an Alternative Typology." *Journal of Politics* 40 (1978):134-161.

Hassoun, Hussein. *Industrial Energy Consumption Studies, Part I. The Aluminum Industry.* Salem, Ore.: Oregon Department of Energy, 1978.

Hassoun, Hussein; Nadal, Joseph; and Timm, Richard. "In the Matter of the Application of Portland General Electric Company for a Site Certificate for the Pebble Springs Nuclear Power Plant." Testimony before the Oregon Energy Facility Siting Council. Salem, Ore.: Oregon Department of Energy, April 28, 1976.

Heclo, Hugh. *A Government of Strangers.* Washington, D.C.: Brookings Institution, 1977.

Hellman, Richard. *Government Competition in the Electric Utility Industry.* New York: Praeger, 1972.

Higgins, Larett. "Electricity and Canadian Policy." In *The Energy Question,* edited by Edward W. Erickson and Leonard Waverman, Vol. 2, pp. 171-88. Toronto: University of Toronto Press, 1974.

Hill, Greg. "The Politics of Energy 1990." Mimeographed. Seattle, Wa.: Institute for Environmental Studies, University of Washington, April 1978.

Hill, Phillip G. *Power Generation.* Cambridge: MIT Press, 1977.

Hinman, George, et al. *Energy Projections for the Pacific Northwest.* Pullman, Wa.: Environmental Research Center, Washington State University, 1975.

Hitch, Charles J. "Energy in Our Future." *Key Reporter.* Phi Beta Kappa newsletter. Summer 1978.

Hittle, Larry. BPA, Office of General Counsel. Interview by KNL, Portland, Ore., July 1979.

Hodel, Donald P. "The Prophets of Shortage." Speech delivered to the City Club of Portland, July 11, 1975. Portland, Ore.: BPA, 1975.

________. "The Prophets of Shortage (Part 2)." Speech before the Rotary Club of Seattle, August 21, 1975. Portland: BPA, 1975.

________. "The Prophets of Shortage (Part 2)." Remarks to the Rotary Club of Seattle, August 27, 1975. Portland, Ore.: BPA, 1975.

________. Interview by KNL. Portland, Ore., February 1978.

Hohenemser, Christoph; Kasperson, Roger; and Kates, Robert. "The Distrust of Nuclear Power." *Science* 196 (1977):25-34.

Houston, Holland H. *The Place of Steam-Electric Generating Stations in the Orderly Program of Electric Power Development in the Pacific Northwest.* Department of Conservation and Development and Public Service Commission, Bulletin 38. Olympia, Wa., 1950.

Hulbert, William G., Jr. Letter to Representative Lloyd Meeds, September 8, 1977. Attachment to "Snohomish Amendment" of H.R. 9020.

Interagency Review Group on Nuclear Waste Management. *Report to the President*, TID-29442, March 1979. Springfield, Va.: National Technical Information Service, 1979.

Ivancie, Francis J. Press statement, Portland Department of Public Utilities, November 12, 1976. Copy on file with DLK.

Joskow, Paul L., and Baughman, Martin L. "The Future of the U.S. Nuclear Energy Industry." *Bell Journal of Economics* 7 (1976):3-32.

Kahn, Alfred E. *The Economics of Regulation*. New York: Wiley, 1970.

Katz, Dean. "Regional Power Bill Finally Surfaces in Senate. Seattle *Times*, July 29, 1978, A8.

Kell, Alterman and Runstein. *Laws Needed to Expedite Electric Generation Schedules for the Pacific Northwest: Financing, Environmental Laws and Siting and Licensing Procedures*. 1976. Portland, Ore.: Copy on file at offices of Kell, Alterman & Runstein, attorneys.

Keohane, Robert O., and Nye, Joseph S. *Power and Interdependence*. Boston: Little, Brown and Co., 1977.

Klemka, Donna Lee. "Pacific Northwest Electrical Energy Planning. Master's thesis, University of Washington, 1978.

Krutilla, John V. *The Columbia River Treaty: Economics of an International River Basin Development*. Baltimore: Johns Hopkins University Press, 1967.

Lave, Charles A., and March, James G. *An Introduction to Models in the Social Sciences*. New York: McGraw-Hill, 1975.

Leach, Richard H. *American Federalism*. New York: Norton, 1970.

Lee, Kai N.; Marts, Marion E.; Walton, Robert G.; and Birmingham, Alison D. "Electric Power Development in the Pacific Northwest Region: Institutional Commitments and Alternatives, Phase 1." Project completion report, Institute for Environmental Studies, University of Washington, September 1976.

Lowi, Theodore J. "American Business, Public Policy, Case Studies, and Political Theory." *World Politics* 16 (July 1964):677-714.

________. *The End of Liberalism*. New York: Norton, 1969.

________. "Decision Making vs. Policy Making: Toward an Antidote for Technocracy." *Public Administration Review* 30 (1970):314-325.

________. "Four Systems of Policy, Politics, and Choice." Ibid., 32 (1972):298-310.

Lukacs, John. *The Passing of the Modern Age*. New York: Harper Torchbooks, 1970.

Marts, Marion E. "When Can California Join the Union?" *Association of Pacific Coast Geographers' Yearbook* 23 (1961): 7-12.

________. "Diverting the Columbia River." *Argus*. Seattle, Wa.: November 21, 1975.

McConnell, Grant. *Private Power and American Democracy.* New York: Vintage, 1966.

McGrath, Tom. Puget Sound Power and Light. Interview by KNL, October 1978.

McKinley, Charles. *Uncle Sam in the Pacific Northwest.* Berkeley: University of California Press, 1952.

Merkel, Joel. Office of Senator Henry M. Jackson, Democrat of Washington State. Interview by KNL. Seattle, Wa., June 1979.

Miller, Perry. "The Responsibility of Mind in a Civilization of Machines." *American Scholar* 31 (Winter 1961-62). Reprinted in *Changing Attitudes Towards American Technology,* edited by Thomas Parke Hughes. New York: Harper and Row, 1975.

Mitchell, Robert Cameron. "The Public Speaks Again: A New Environmental Survey." *Resources* 60 (September-November 1978):1-6. Newsletter of Resources for the Future, Washington, D.C.

Morison, Elting E. *From Know-How to Nowhere.* New York: Basic Books, 1974.

Mosey, Ed. "Governors Split Over Power Plan." Portland *Oregonian,* April 12, 1978.

Myhra, David. "One Nuke Gets a Warm Welcome." *Planning* (November 1975):13-18.

Nagel, Theodore, J. "Operating a Major Electric Utility Today." *Science* 201 (1978):985-993.

Nelkin, Dorothy, and Fallows, Susan. "The Evolution of the Nuclear Debate: The Role of Public Participation." *Annual Review of Energy* 3 (1978):275-312.

Noll, Roger G. *Reforming Regulation.* Washington, D.C.: Brookings Institution, 1971.

Northern Plains Resource Council. "Future Northwest Power Needs Pose Serious Threat to Plains States." *Washington Resource Report* (April 1978):4-5. Newsletter.

Northwest Energy Policy Project. *Energy Futures Northwest.* Springfield, Va.: National Technical Information Service, 1978.

Northwest Environmental Technology Laboratories. *Energy 1990 Consultants' Report.* Bellevue, Wa.: NETL, 1976.

Nuclear Energy Policy Study Group. *Nuclear Power Issues and Choices.* Cambridge, Mass.: Ballinger, 1977.

O'Connor, James. *The Fiscal Crisis of the State.* New York: St. Martin's Press, 1973.

Olson, Mancur. *The Logic of Collective Action.* Cambridge: Harvard University Press, 1965.

Oregon Department of Energy. *Oregon's Energy Future--First Annual Report.* Salem, Ore.: ODE, 1977.

_______. *Oregon's Energy Future--Second Annual Report.* Salem, Ore.: ODE, 1978.

Oregon State Legislative Assembly. S.B. 320. "The DRPA bill."

Owen, Marguerite. *The Tennessee Valley Authority*. New York: Praeger, 1973.

Pacific Northwest River Basins Commission. *Review of Power Planning in the Pacific Northwest, Calendar Year 1975*. [Vancouver, Wa.]: PNRBC (1976).

_______. *Review of Power Planning in the Pacific Northwest--Calendar Year 1976*. [Vancouver, Wa.]: PNRBC (1977).

_______. *Review of Power Planning in the Pacific Northwest--Calendar Year 1977*. [Vancouver, Wa.]: PNRBC (1978).

_______, Power Planning Committee. *Power Planning Organizations in the Pacific Northwest*. [Vancouver, Wa.], (1978).

Pacific Northwest Utilities Conference Committee. *Long Range Projection of Loads and Resources 1977-78 through 1996-97, West Group Area*. N.p.: PNUCC, April 1977.

Pendergrass, Bonnie Baack. "Public Power, Politics, and Technology in the Eisenhower and Kennedy Years: The Hanford Dual-Purpose Reactor Controversy, 1956-1962." Ph.D. dissertation, University of Washington, 1974.

Perrow, Charles. *Complex Organizations: A Critical Essay*. Glenview, Ill.: Scott Foresman, 1972.

Peterson, Paul E. "Forms of Representation: Participation of the Poor in the Community Action Program." *American Political Science Review* 64 (1970):491-507.

Piven, Frances Fox, and Cloward, Richard C. *Regulating the Poor*. New York: Vintage, 1971.

PNUCC Task Force 8. "Thermal Plant Operating Characteristics." Report to the Technical Coordinating Group. N.p.: PNUCC, September 20, 1977.

Price, Don K. *The Scientific Estate*. New York: Oxford University Press, 1968.

Profozich, Russell J. "The Changing Structure of the Municipal Utility Sector." Report No. 78-219E, Congressional Research Service. Washington, D.C.: Library of Congress, November 6, 1978.

Pugnetti, Donald A. "Jackson Offer on Hanford Steam Plant Is Studied." Pasco, Wa.: *Tri-City Herald*, July 12, 1962.

Redman, Eric. Attorney for direct-service industrial customers. Private communication to KNL, July 1978; November 1978.

_______. Letter to Joan Hohl and KNL, with attachments, July 14, 1978.

Reich, George. Seattle area office of BPA. Conversation with KNL, Seattle, Wa., November 1978.

Ricco, Edward. "Anti-Nuclear Sentiment Makes Itself Felt." *Natural Resources Journal* 19 (1979):377-379.

Rochlin, Gene I. *Plutonium, Power, and Politics*. Berkeley: University of California Press, 1979.

Rodgers, William H., Jr. "Siting Power Plants in Washington State." *Washington Law Review* 47 (1971):10n.

Ross, William M., and Marts, Marion E. "The High Ross Dam Project." *Canadian Geographer* 19 (1975):221-234.
Rourke, Francis E. *Bureaucracy, Politics, and Public Policy.* Boston: Little Brown and Co., 1969.
Ruggie, John Gerard. "International Responses to Technology: Concepts and Trends." *International Organization* 29 (1975): 557-583.
Salisbury, Robert H. "The Analysis of Public Policy: A Search for Theories and Roles." In *Political Science and Public Policy,* edited by Austin Ranney. Chicago: Markham, 1968.
Schultze, Charles L. *The Public Use of Private Interest.* Washington, D.C.: Brookings Institution, 1977.
Scheingold, Stuart A. *The Politics of Rights.* New Haven: Yale University Press, 1974.
________. "The Community in Perspective: Public Policy and Political Structure." *Annals American Academy of Political and Social Sciences* 440 (1978):156-167.
Seligman, Dan. "The Washington Public Power Supply System." Mimeographed. Seattle: Institute for Environmental Studies, University of Washington, 1978.
Selznick, Philip. *TVA and the Grass Roots.* Berkeley: University of California Press, 1949.
________. *Leadership in Administration.* New York: Harper and Row, 1957.
Shapley, Deborah. "TVA Today: Former Reformers in an Era of Expensive Electricity." *Science* 194 (1976):814-818.
________. "Nuclear Power Plants: Why Do Some Work Better Than Others?" *Science* 195 (1977):1311-1313.
Socolow, Robert H. "Failures of Discourse." In *Boundaries of Analysis,* edited by Harold A. Feiverson; Frank W. Sinden; and Socolow. Cambridge, Mass.: Ballinger, 1976.
Speth, J. Gustave. Speech to the American Law Institute/American Bar Association Environmental Law Conference, September 29, 1977, Washington, D.C. Reported in *Environmental Reporter* "Current Developments 8:23 (October 7, 1977):855.
Springer, Vera. *Power and the Pacific Northwest.* Portland, Ore.: BPA [1976].
Stein, Herbert A. *The Fiscal Revolution in America.* Chicago: University of Chicago Press, 1968.
Stoler, Peter. "The Irrational Fight Against Nuclear." *Time* (September 25, 1978):71-72.
Stoll, Norman A. "Pros and Cons of Proposed Regional Power Supply Program as Developed to Date." Memorandum to Public Power Council Executive Committee, April 22, 1977, p. 13. Copy on file with DLK.
Swainson, Neil A. *Conflict Over the Columbia.* Montreal: McGill-Queen's University Press, 1979.
Swift, Al. Democratic Representative for Washington State. Interview by KNL, Washington, D.C., June 1979.

Taylor, J. Serge. "Environmentalists Inside the Bureaucracy: Environmental Impact Analysis in the Forest Service and Army Corps of Engineers." Ph.D. dissertation, University of California, 1978.

_______. Letter to KNL, September 1978.

Tilly, Charles. *From Mobilization to Revolution.* Reading, Mass.: Addison-Wesley, 1978.

Timm, Richard, and Nadal, Joseph. Testimony on proposed specific standards to Oregon Energy Facility Siting Council. Mimeographed. Salem, Ore.: Oregon Department of Energy, September 22, 1977.

Thomas, Robert D. "Intergovernmental Coordination in the Implementation of National Air and Water Pollution Policies." In *Public Policy Making in a Federal System,* edited by Charles O. Jones and Robert D. Thomas, Chapter 6. Beverly Hills, Calif.: Sage, 1976.

Thompson, James D. *Organizations in Action.* New York: McGraw-Hill, 1967.

Tucker, William. "Environmentalism and the Leisure Class." *Harpers* (December 1977):49-80.

Tupper, George, and Noguchi, Tom. Seattle area office of BPA. Interview by DLK. Seattle, May 23, 1978.

U.S. Army Corps of Engineers. *Review Report on Columbia River and Tributaries.* House Document 403, 87th Congress, 2nd Session, 1958. This is updated version of House Document 531, 81st Congress, 2nd Session, 1948.

_______, North Pacific Division. "Pumped-Storage Inventory of the Pacific Northwest." Prepared as an element of the Columbia River and Tributaries Review Study. Report No. 26. Portland: January 1976. Copy on file in the office of Professor Marion E. Marts, University of Washington, Seattle.

U.S. Code, Vol. 16, Chap. 12B (1964 ed.). Bonneville Project Act, Act of August 20, 1937, Ch. 720, 50 Stat. 731, as amended.

U.S. Comptroller General. *Impacts and Implications of the Pacific Northwest Power Bill,* EMD-79-105. Washington, D.C.: U.S. General Accounting Office, September 4, 1979.

_______. *Region at the Crossroads--the Pacific Northwest Search for New Sources of Electric Energy,* EMD-78-76. Washington, D.C.: U.S. General Accounting Office, August 10, 1978.

U.S. Congress. House of Representatives. *Pacific Northwest Electric Power Supply and Conservation,* hearings before the Subcommittee on Water and Power, Committee on Interior and Insular Affairs, 95th Congr., 1st Sess., Serial No. 95-26, in five parts. Washington, D.C.: Committee Print, 1978.

_______. *Hearings on Pacific Northwest Electric Power Issues,* before Committee on Interstate and Foreigh Commerce, Subcommittee on Energy and Power. 95th Congr., 2d Sess., Serial No. 95-193. Washington, D.C.: Committee Print, 1979.

_______. Senate. *Hearings on Pacific Northwest Electric Power Supply and Conservation Act,* before Committee on Energy and Natural Resources. 95th Congr., 2d Sess. Publication No. 95-154, in two parts. Washington, D.C.: Committee Print, 1979.

_______. *Pacific Northwest Electric Power Planning and Conservation Act,* report No. 96-272, 96th Congr., 1st Sess., July 30, 1979. (To accompany S. 885).

_______. Legislation.

S. 2080. "Pacific Northwest Electric Supply and Conservation Act." 95th Congr., 1st Sess., September 1977.

H.R. 9020. "Pacific Northwest Electric Supply and Conservation Act." 95th Congr., 1st Sess., September 1977.

H.R. 5862. "Columbia Basin Energy Corporation Act of 1977." 95th Congr., 1st Sess., March 1977.

S. 3418. "Pacific Northwest Electric Power Planning and Conservation Act." 95th Congr., 2d Sess., August 1978.

H.R. 13931. "Pacific Northwest Electric Power Planning and Conservation Act." 95th Congr., 2d Sess., August 1978.

S. 885. "Pacific Northwest Electric Power Planning and Conservation Act." 96th Congr., 1st Sess., April 1979.

H.R. 3508. "Pacific Northwest Electric Power Planning and Conservation Act." 96th Congr., 1st Sess., Fall 1979.

H.R. 4159. "Columbia Basin Energy Corporation Act of 1978." 96th Congr., 1st Sess, May 1979.

U.S. Department of Energy. *The National Energy Act.* DOE - OPA-0003. U.S. Washington, D.C.: U.S. Government Printing Office, 1978.

U.S. Energy Research and Development Administration. *Alternatives for Managing Wastes from Reactors and Post-Fission Operations in the LWR Fuel Cycle,* ERDA 76-43. Springfield, Va.: National Technical Information Service, 1976.

U.S. Department of Interior, Heritage Conservation and Recreation Service. *Impacts on Recreation from Hydroelectric Power Peaking on the Mid-Columbia River, Washington.* Seattle, Wa.: Northwest Regional Office, July 1978.

U.S. Nuclear Regulatory Commission. *Nuclear Energy Center Site Survey,* report NUREG 76-0001. Springfield, Va.: National Technical Information Service, 1976.

Wagner, Steven K. "NW Governors Clash With Ray Over Power." Seattle *Post-Intelligencer,* April 12, 1978.

Walsh, John. "Opposition to Nuclear Power: Raising the Question at the Polls." *Science* 190 (1975):964-966.

_______. "Fuel Reprocessing Still the Focus of U.S. Nonproliferation Policy. *Science* 201 (1978):692-697.

Walton, Robert G. "The Development of the Electrical Power System in the Pacific Northwest, A Public Policy Perspective." Master's thesis, University of Washington, 1976.

Washington Public Interest Research Group. Draft study. On file in the office of KNL, Institute for Environmental Studies, University of Washington, Seattle, Wa.: July 14, 1978.

Washington Public Power Supply System. 1977 Annual Report. Richland, Wa.: WPPSS, 1978.

_______. *1980 Budget*. Richland, Wa.: WPPSS, 1977.

Washington State Power Commission. "Steam-Electric Generating Facilities," embodying study done by firm of Ford, Bacon & Davis. Preliminary report. Olympia, Wa.: November 23, 1956. Copy on file at office of Professor Marion E. Marts, University of Washington, Seattle.

Water Resources Policy Commission. *Water Policy for the American People*. "Report of the President's Water Resources Policy Commission." Washington, D.C.: U.S. Government Printing Office, 1950. Also: *Resources Law*, Vol. 3, pp. 281-84, 298-99.

Whittlesey, Norman K., et al. *Benefits and Costs of Irrigation Development in Washington*. Report prepared for the House of Representatives and the Legislative Transportation Committee, Washington State Legislature, October 1976. Vol. 1, p. 20. Olympia, Wa.

Wiebe, Robert. *The Search for Order, 1877-1920*. New York: Hill and Wang, 1967.

Williams, Roger M. "Oregon: The Fight for 'Survival'." *Saturday Review/World* (November 16, 1974):10-15.

Willrich, Mason. "The Electric Utility and the Energy Crisis." *Public Utilities Fortnightly* (January 2, 1975):22-28; and ibid., (January 16, 1975):25-34.

Wilson, James Q. *Political Organizations*. New York: Basic Books, 1973.

Wilson, Woodrow. *Congressional Government*. Boston: Houghton Mifflin, Co., 1885.

Winner, Langdon. *Autonomous Technology*. Cambridge, Mass.: MIT Press, 1977.

Wolin, Sheldon. *Politics and Vision*. Boston: Little, Brown and Co., 1960.

Wood, Robert C. "When Government Works." *Public Interest*, No. 18 (Winter 1970):39-51.

Woodfill, Douglas. "Pacific Northwest Power Entities, An Expanded Glossary." Mimeographed. Seattle, Wa.: Seattle City Light, December 1975.

Index